SpringerBriefs in Applied Sciences and Technology

Series editor

Janusz Kacprzyk, Polish Academy of Sciences, Systems Research Institute,
Warsaw, Poland

SpringerBriefs present concise summaries of cutting-edge research and practical applications across a wide spectrum of fields. Featuring compact volumes of 50–125 pages, the series covers a range of content from professional to academic.

Typical publications can be:

- A timely report of state-of-the art methods
- An introduction to or a manual for the application of mathematical or computer techniques
- A bridge between new research results, as published in journal articles
- A snapshot of a hot or emerging topic
- An in-depth case study
- A presentation of core concepts that students must understand in order to make independent contributions

SpringerBriefs are characterized by fast, global electronic dissemination, standard publishing contracts, standardized manuscript preparation and formatting guidelines, and expedited production schedules.

On the one hand, **SpringerBriefs in Applied Sciences and Technology** are devoted to the publication of fundamentals and applications within the different classical engineering disciplines as well as in interdisciplinary fields that recently emerged between these areas. On the other hand, as the boundary separating fundamental research and applied technology is more and more dissolving, this series is particularly open to trans-disciplinary topics between fundamental science and engineering.

Indexed by EI-Compendex and Springerlink.

More information about this series at http://www.springer.com/series/8884

Shuai Li · Long Jin

Competition-Based Neural Networks with Robotic Applications

Shuai Li
The Hong Kong Polytechnic University
Hong Kong
China

Long Jin
Lanzhou University
Lanzhou
China

ISSN 2191-530X   ISSN 2191-5318   (electronic)
SpringerBriefs in Applied Sciences and Technology
ISBN 978-981-10-4946-0   ISBN 978-981-10-4947-7   (eBook)
DOI 10.1007/978-981-10-4947-7

Library of Congress Control Number: 2017941544

Printed on acid-free paper

This Springer imprint is published by Springer Nature
The registered company is Springer Nature Singapore Pte Ltd.
The registered company address is: 152 Beach Road, #21-01/04 Gateway East, Singapore 189721, Singapore

To our ancestors and parents, as always

Preface

In past years, dynamic consensus has attracted intensive research attentions and led to successful solutions of a large variety of distributed computation problems, including distributed formation control, distributed Kalman filter, decentralized control of swarm statistics, and synchronization of networked oscillators. Despite its great success, consensus algorithm, which updates the state by dynamically mitigating differences among agents, is mostly limited to the modeling of dynamic cooperation. It essentially lacks a mechanism to model dynamic competition in a distributed network, which desires the increase of peer differences and the enhancement of contrasts.

Research in many fields confirms the same importance of competition as that of cooperation in the emergence of complex behaviors. For example, it is revealed that competition and cooperation plays significant roles in the decision making in market economy and that the strategy chosen by the rational politicians consists of cooperation over competition in dealing with international relationships. Recent research in neuroscience found finds that control actions depend on transitory change in patterns of cooperation and competition between brain systems during cognitive control. Due to the fundamental significance of competition in the interaction of multi-agent systems, various models have been presented to capture this competitive nature. Among them, the winner-take-all (WTA) model, which refers to the competition of a group of agents that the one with the largest input finally remains activated, while all the other agents are deactivated, has been widely investigated and usually employed to model competition behaviors. Maass proves that a two-layered network composed of weighted averaging in the first layer and WTA in the second layer is able to approximate any nonlinear mapping in any desired accuracy. Following this results, the dynamic consensus with the capability for the computation of weighted averaging in a distributed way, and a distributed algorithm for the computation of WTA, will be able to constitute any nonlinear mapping in a distributed network.

In the past two decades, recurrent neural networks have received considerable studies in many scientific and engineering fields. Particularly, after the invention of the well-known Hopfield neural network, which was originally designed for real-time optimization, the recurrent neural network, as a powerful online optimization tool with potential parallel implementations, is becoming an independent research direction in online optimization field. Remarkable advances have been made in the area of recurrent neural networks for online optimization. To a constrained optimization problem, early works often remove the explicit constraints by introducing a penalty term into the cost function and then design a recurrent neural network evolving along the gradient descent direction. This type of neural networks only converges to an approximation of the optimal solution. In order to obtain a recurrent neural network with guaranteed convergence to the optimal solution, later works introduce dynamic Lagrange multipliers to regulate the constraints. There exist various mathematical models for the description of the WTA competition. By following optimization-based formulation, WTA problem can be modeled as a constrained convex quadratic programming (QP) problem, and then, traditionally, gradient descent or projected gradient descent is employed to get the corresponding dynamic equations for online solution of the problem.

In this book, focusing on solving competition-based problems, we design, propose, develop, analyze, model, and simulate various neural network models depicted in centralized and distributed manners. Specifically, we define four different classes of centralized models for investigating the resultant competition in a group of multiple agents. For distributed competition with limited communication among agents, we present the first distributed WTA protocol and then extend it to the distributed coordination control of multiple robots. As for these models, the related theoretical analyses are given, and the corresponding modeling is illustrated. Computer simulations with various illustrative examples are performed to substantiate the efficacy of the proposed recurrent neural network models for solving WTA problems. Based on these successful researches, we further apply such a distributed WTA approach to distributed coordination control of multiple redundant robot manipulators. The corresponding results show the application prospect of the presented competition-based neural network approach to robot applications.

The idea for this book on solving competition-based problems was conceived during the classroom teaching as well as the research discussion in the laboratory and at international scientific meetings. All of the materials of this book are derived from the authors' papers published in journals, such as IEEE Transactions on Automatic Control, IEEE Transactions on Neural Networks and Learning Systems, IEEE Transactions on Systems, Man, and Cybernetics: Systems, Neural Networks. In fact, since the early 1980s, the field of neural networks has undergone the phases of exponential growth, generating many new theoretical concepts and tools (including the authors' ones). At the same time, these theoretical results have been applied successfully to the solution of many practical problems. Our first priority is thus to cover each central topic in enough details to make the material clear and coherent; in other words, each part (and even each chapter) is written in a relatively self-contained manner.

This book is classified into the following 6 chapters.

Chapter 1—In this chapter, we investigates a simple discrete-time model, which produces the winner-take-all competition. The local stability and global stability of the model are both proven theoretically. Simulations are conducted for both the static competition and the dynamic competition scenarios. The numerical results validate the theoretical results and demonstrate the effectiveness of the model in generating winner-take-all competition.

Chapter 2—In this chapter, different from the model presented in Chap. 1, we present a continuous-time dynamic model, which is described by an ordinary differential equation and is able to produce the winner-take-all competition by taking advantage of selective positive–negative feedback. The global convergence is proven analytically, and the convergence rate is also discussed. Simulations are conducted in the static competition and the dynamic competition scenarios. Both theoretical and numerical results validate the effectiveness of the dynamic equation in describing the nonlinear phenomena of winner-take-all competition.

Chapter 3—In this chapter, a class of recurrent neural networks to solve quadratic programming problems are presented and further extended to competition generation. Different from most existing recurrent neural networks for solving quadratic programming problems, the proposed neural network model converges in finite time and the activation function is not required to be a hard-limiting function for finite convergence time. The stability, finite-time convergence property, and the optimality of the proposed neural network for solving the original quadratic programming problem are proven in theory. Extensive simulations are performed to evaluate the performance of the neural network with different parameters. In addition, the proposed neural network is applied to solving the k-winner-take-all (k-WTA) problem. Both theoretical analysis and numerical simulations validate the effectiveness of our method for solving the k-WTA problem.

Chapter 4—In this chapter, we make steps in that direction and present a simple model, which produces the winner-take-all competition by taking advantage of selective positive–negative feedback through the interaction of neurons via p-norm. Compared to models presented in Chaps. 1–3, this model has an explicit explanation of the competition mechanism. The ultimate convergence behavior of this model is proven analytically. The convergence rate is also discussed. Simulations are conducted in the static competition and the dynamic competition scenarios. Both theoretical and numerical results validate the effectiveness of the dynamic equation in describing the nonlinear phenomena of winner-take-all competition.

Chapter 5—When it comes to distributed networks, Maass's theorem poses great appeal for distributed WTA algorithms provided that the distributed weighted averaging could be addressed using consensus. Unfortunately, as presented in Chap. 1 through Chap. 4, there is no existing distributed WTA algorithm available, which significantly blocks the exhibition of the computational power of WTA over dynamic networks. In this chapter, we make progress along this direction and present the first distributed WTA protocol with guaranteed global convergence. The convergence to the WTA solution is proved rigourously using Lyapunov theory. The theoretical conclusions are supported by numerical validation.

Chapter 6—In this chapter, as an application of the competition-based models investigated in previous chapters, the problem of dynamic task allocation in a distributed network of redundant robot manipulators for path-tracking with limited communications is investigated, where k fittest ones in a group of n redundant robot manipulators with $n > k$ are allocated to execute an object tracking task. The problem is essentially challenging in view of the interplay of manipulator kinematics and the dynamic competition for activation among manipulators. To handle such an intricate problem, a distributed coordination control law is developed for the dynamic task allocation among multiple redundant robot manipulators with limited communications and with the aid of a consensus filter. In addition, a theorem and its proof are presented for guaranteeing the convergence and stability of the proposed distributed control law. Finally, an illustrative example is provided and analyzed to substantiate the efficacy of the proposed control law.

In summary, this book presents models producing the WTA competition in centralized and distributed manners and further applies these models to distributed coordination control of multiple robot manipulators (showing its application prospect). This book is written for graduate students as well as academic and industrial researchers studying in the developing fields of neural dynamics, computer mathematics, time-varying computation, simulation and modeling, analog hardware, and robotics. It provides a comprehensive view of the combined research of these fields, in addition to its accomplishments, potentials, and perspectives. We do hope that this book will generate curiosity and also happiness to its readers for learning more in the fields and the research and that it will provide new challenges to seek new theoretical tools and practical applications.

At the end of this preface, it is worth pointing out that, in this book, a new and inspiring direction on the competition-based neural network as well as its applications is provided. This opens an avenue to study distributed competition over a connected network with any possible topology. It may promise to become a major inspiration for studies and researches in neural dynamics, robotics, and dynamic decision making. Without doubt, this book can be extended. Any comments or suggestions are welcome. The authors can be contacted via e-mails: shuaili@polyu.edu.hk and jinlong@lzu.edu.cn.

Hong Kong, China   Shuai Li
Lanzhou, China   Long Jin
March 2017

Acknowledgements

During the work on this book, we have had the pleasure of discussing its various aspects and results with many cooperators and students. We highly appreciate their contributions, which particularly allowed us to improve our book manuscript.

The continuous support of our research by the National Natural Science Foundation of China (with number 61401385), by Hong Kong Research Grants Council Early Career Scheme (with number 25214015), and also by Departmental General Research Fund of Hong Kong Polytechnic University (with number G.61.37.UA7L) is gratefully acknowledged here either.

Besides, we would like to thank the editors sincerely for their very important and constructive comments and suggestions provided, in addition to their time and efforts spent in handling this book.

We are always very grateful to the nice people (especially the staff in Springer) for their strong support and encouragement during the preparation and publishing of this book.

Contents

Chapter 1
Competition Aided with Discrete-Time Dynamic Feedback

Abstract In this chapter, we investigates a simple discrete-time model, which produces the winner-take-all competition. The local and global stability of the model are both proven theoretically. Simulations are conducted for both the static competition and the dynamic competition scenarios. The numerical results validate the theoretical results and demonstrate the effectiveness of the model in generating winner-take-all competition.

Keywords Winner-take-all competition · Dynamic feedback · Discrete-time system · Global stability · Discrete-time competition-based neural networks.

1.1 Introduction

Winner-take-all refers to the phenomena that agents in a group compete with each others for activation and only the one with the highest input stays active while all the others deactivated. The winner-take-all models many competition phenomena existing in nature [1, 2] and finds applications in many engineering fields [3]. It is remarkably that the winner-take-all competition is computationally powerful and can generate some useful functions required in computational intelligence applications [3]. Due to the importance of winner-take-all competition in engineering applications, there have been many attempts to design circuits for its implementation [4, 5].

There have been various models presented by researchers to explain or generate the winner-take-all behavior. In [6], the N species Lotka–Volterra model is used for the explanation. Inspired by the great success of recurrent neural networks [7–22], recurrent neural networks are utilized to investigate the winner-take-all competition [23, 24]. In [25, 26], the FitzHugh–Nagumo Model, which is able to demonstrate the interactive spiking, is used to study the winner-take-all behavior. In [27, 28], the winner-take-all problem is regarded as the solution of an optimization problem and the result is generated by solving such a problem. Although many mathematic models have been presented to explain or generate the winner-take-all competition, it is still an open problem to find a simple model describing such a nonlinear phenomena with rigorous analysis on its performance. Neural networks, which possess the abilities of high-speed parallel distributed processing, and can be implemented by hardware,

© The Author(s) 2018

S. Li and L. Jin, *Competition-Based Neural Networks with Robotic Applications*,
SpringerBriefs in Applied Sciences and Technology,
DOI 10.1007/978-981-10-4947-7_1

have been recognized as a powerful tool for real-time processing and successfully applied widely in various control systems [29–34]. In this chapter, we present a simple model described by nonlinear difference equations to generate the winner-take-all competition. The solution of the equilibrium points, the local stability and the global stability are resolved in theory. Due to the simplicity of this model, it has strong potentials to be implemented in hardware with less hardware complexity compared with some existing models.

1.2   Problem Definition

In this section, we define the winner-take-all competition in the following way: the agent with the largest input finally wins the competition and keeps activated while all the other agents withe smaller inputs are deactivated to zero eventually.

This definition is a mathematical abstraction of many competition phenomena found in nature and society, such as the growth competition in plants [1], competitive decision making in the cortex [35], foraging and mating in animal societies [2].

1.3   Model Formulation

The presented model has the following dynamic for the ith agent in a group of totally n agents,

$$
\begin{aligned}
x_{1i}(t+1) &= u_i x_{2i}(t), \\
x_{2i}(t+1) &= \frac{x_{1i}(t+1)}{\|x_1(t+1)\|}, \\
y_i(t+1) &= x_{2i}(t+1),
\end{aligned}
\tag{1.1}
$$

where t represent time instant, $u_i \in \mathbb{R}$ is the input and $u_i \geq 0$, $u_i \neq u_j$ for $i \neq j$, $x_1(t) = [x_{11}(t), x_{12}(t), \ldots, x_{1n}(t)]^{\mathsf{T}} \in \mathbb{R}^n$, $x_{1i}(t) \in \mathbb{R}$ and $x_{2i}(t) \in \mathbb{R}$ for $i = 1, 2, \ldots, n$ denote the first and the second state value of the ith agent at time t respectively, $\|x_1(t)\| = \sqrt{x_{11}^2(t) + x_{12}^2(t) + \cdots + x_{1n}^2(t)}$ denotes the Euclidean norm of $x_1(t)$, $y_i(t) \in \mathbb{R}$ represents the output of the ith agent at time t.

The dynamic equation (1.1) can be written into the following compact form by stacking up the state for all agents,

$$
\begin{aligned}
x_1(t+1) &= u \circ x_2(t), \\
x_2(t+1) &= \frac{x_1(t+1)}{\|x_1(t+1)\|}, \\
y(t+1) &= x_2(t+1),
\end{aligned}
\tag{1.2}
$$

where $u = [u_1, u_2, \ldots, u_n]^T \in \mathbb{R}^n$, $x_1(t) = [x_{11}(t), x_{12}(t), \ldots, x_{1n}(t)]^T \in \mathbb{R}^n$, $x_2(t) = [x_{21}(t), x_{22}(t), \ldots, x_{2n}(t)]^T \in \mathbb{R}^n$, $y(t) = [y_1(t), y_2(t), \ldots, y_n(t)]^T \in \mathbb{R}^n$, the operator 'o' represents the multiplication in component-wise, i.e., $u \circ x = [u_1 x_1, u_2 x_1, \ldots, u_n x_n]^T$.

1.4 Theoretical Results

In this section, theoretical results on the dynamic system (1.1) are presented. We first examine the equilibrium point of the dynamic system and then investigate its local stability around the equilibria. After that, we turn to the proof of the global stability of the system.

On the equilibrium points of the dynamic system (1.1), we have the following theorem,

Theorem 1.1 *The discrete-time dynamic system (1.1) has equilibrium points at* $(x_1^*, x_2^*, y^*) = \pm(u_i e_i, e_i, e_i)$ *for* $i = 1, 2, \ldots, n$, *where* $u_i \geq 0$ *is the ith input,* $e_i \in \mathbb{R}^n$ *is a n dimensional vector with the ith element equal 1 and all the others equal zero.*

Proof In Eq. (1.2), letting $x_1(t + 1) = x_1^*$, $x_2(t + 1) = x_2(t) = x_2^*$, $y(t + 1) = y^*$, we get the following equations for the equilibrium points,

$$x_1^* = u \circ x_2^*, \tag{1.3a}$$

$$x_2^* = \frac{x_1^*}{\|x_1^*\|}, \tag{1.3b}$$

$$y^* = x_2^*. \tag{1.3c}$$

According to the definition of the operator 'o'. Equation (1.3a) can be written as

$$x_1^* = \text{diag}(u)x_2^*, \tag{1.4}$$

where $\text{diag}(u)$ is defined as the matrix with u as the diagonal elements and all the other elements zero. Together with Eq. (1.3b), we have,

$$x_1^* = \text{diag}(u)\frac{x_1^*}{\|x_1^*\|}, \tag{1.5}$$

i.e.,

$$\|x_1^*\|x_1^* = \text{diag}(u)x_1^*. \tag{1.6}$$

Clearly, this is an eigen-equation for the matrix $\text{diag}(u)$. The existence of the solution requires x_1^* being the eigenvector of $\text{diag}(u)$ and $\|x_1^*\|$ being the corresponding eigenvalue. As the matrix $\text{diag}(u)$ is diagonal, its eigenvalue and normalized eigenvector pairs can be easily got as u_1 with e_1, or u_2 with e_2, or, u_3 with e_3, ..., or u_n with e_n, respectively. Comparing the eigenvalue and eigenvector pairs of diag(u) with (1.6), we get the solution of x_1^* as: $x_1^* = \pm u_1 e_1$, $x_1^* = \pm u_2 e_2$, ..., or $x_1^* = \pm u_n e_n$. From (1.3b) and (1.3c), it can be observed that both x_2^* and y^* equal the normalized vector of x_1, i.e., the corresponding solutions of x_2^* are $e_1, e_2, ...,$ or e_n and y^* takes the same solution. To summarize, the equilibrium points are $(x_1^*, x_2^*, y^*) = \pm(u_i e_i, e_i, e_i)$ for $i = 1, 2, ..., n$. This completes the proof.

We have the following results on the local stability of the system (1.1),

Theorem 1.2 *Point* $(x_1^*, x_2^*, y^*) = \pm(u_j e_j, e_j, e_j)$ *is an unstable equilibrium point of the discrete-time dynamic system (1.1) for* $j = 1, 2, ..., n$, $j \neq k^*$, *where* $k^* = argmax_{i=1,2,...,n}(u_i)$, $u_j \geq 0$ *is the jth input,* $e_j \in \mathbb{R}^n$ *is a n dimensional vector with the jth element equal 1 and all the others equal zero.*

Proof Without losing generality, we only consider the equilibrium points $(x_1^*, x_2^*, y^*) = (u_j e_j, e_j, e_j)$ in this proof. For the rest equilibrium points $(x_1^*, x_2^*, y^*) = -(u_j e_j, e_j, e_j)$, the local stability can be analyzed in the same way.

The system (1.2) is a nonlinear difference equation due to the presence of the normalization operation. We use the liberalization technique to analyze the local stability. According to Theorem 1.2, there are totally n equilibrium points for the dynamic system (1.1) and the jth one is $(x_1^*, x_2^*, y^*) = (u_j e_j, e_j, e_j)$. From (1.2), we get the dynamics of x_2 as follows,

$$x_2(t + 1) = \frac{\text{diag}(u)x_2(t)}{\|\text{diag}(u)x_2(t)\|}. \tag{1.7}$$

At the equilibrium point $(x_1^*, x_2^*, y^*) = (u_j e_j, e_j, e_j)$, we have the fact that $\|\text{diag}(u)x_2^*\| = \|\text{diag}(u)e_j\| = \|u_j e_j\| = |u_j| = u_j$. Accordingly, we have the following approximate dynamics around this equilibrium point,

$$x_2(t + 1) = \frac{\text{diag}(u)x_2(t)}{u_j}. \tag{1.8}$$

This is a linear system with $\frac{1}{u_j}\text{diag}(u)$ as the system matrix. $\frac{1}{u_j}\text{diag}(u)$ is a diagonal matrix and thus its eigenvalues are $\frac{u_1}{u_j}, \frac{u_2}{u_j}, ..., \frac{u_n}{u_j}$. For $j \neq k^*$, we have, $u_j < u_{k^*}$ according to the definition of k^*. The k^*th eigenvalue of (1.8), which is $\frac{u_{k^*}}{u_j} > 1$. The linear system (1.8) is unstable since its system matrix has an eigenvalue outside the unit circle. Therefore the nonlinear system with (1.8) as its linear approximation is also unstable. Thus, we conclude that the system (1.1) is unstable at $(x_1^*, x_2^*, y^*) = (u_j e_j, e_j, e_j)$ for $j = 1, 2, ..., n$, $j \neq k^*$. Following the same procedure, we can also conclude that the system (1.1) is unstable at $(x_1^*, x_2^*, y^*) = -(u_j e_j, e_j, e_j)$ for $j = 1, 2, ..., n$, $j \neq k^*$. This completes the proof.

The global stability results on the system (1.1) are stated as follows,

Theorem 1.3 *For any random initializations, the output y_i for $i = 1, 2, \ldots, n$ of the discrete-time dynamic system (1.1) converges to 1 for $i = k^*$ when $x_{2i}(0) > 0$, converges to -1 for $i = k^*$ when $x_{2i}(0) < 0$ and converges to 0 for other i s with $x_{2i}(0) < x_{2k^*}$, where k^* defines the label of the winner, i.e., $k^* = \text{argmax}_{i=1,2,\ldots,n}(u_i)$.*

Proof From (1.2), we get the dynamics of x_2 as follows,

$$x_2(t+1) = \frac{\text{diag}(u)x_2(t)}{\|\text{diag}(u)x_2(t)\|} \tag{1.9}$$

By iteration, we have,

$$\begin{aligned}
x_2(t+1) &= \frac{\text{diag}(u)x_2(t)}{\|\text{diag}(u)x_2(t)\|}, \\
&= \frac{\text{diag}^2(u)x_2(t-1)}{\|\text{diag}^2(u)x_2(t-1)\|}, \\
&\cdots \\
&= \frac{\text{diag}^{t+1}(u)x_2(0)}{\|\text{diag}^{t+1}(u)x_2(0)\|}, \\
&= \frac{\left(\frac{1}{u_{k^*}}\text{diag}(u)\right)^{t+1}x_2(0)}{\|\left(\frac{1}{u_{k^*}}\text{diag}(u)\right)^{t+1}x_2(0)\|}, \\
&= \frac{\text{diag}(\frac{u}{u_{k^*}})^{t+1}x_2(0)}{\|\text{diag}(\frac{u}{u_{k^*}})^{t+1}x_2(0)\|}. \tag{1.10}
\end{aligned}$$

Note that $\text{diag}\left(\frac{u}{u_{k^*}}\right)$ is a diagonal matrix with the ith diagonal element being $\frac{u_i}{u_{k^*}}$ with $|\frac{u_i}{u_{k^*}}| < 1$ for $i \neq k^*$ and $|\frac{u_i}{u_{k^*}}| = 1$ for $i = k^*$. Accordingly, we can compute that $\text{diag}\left(\frac{u}{u_{k^*}}\right)^{t+1}$ is a diagonal matrix with the ith diagonal element equal to $\left(\frac{u_i}{u_{k^*}}\right)^{t+1}$ and we obtain that $\lim_{t\to\infty}\left(\frac{u_i}{u_{k^*}}\right)^{t+1} = 0$ for $i \neq k^*$ and $\lim_{t\to\infty}\left(\frac{u_i}{u_{k^*}}\right)^{t+1} = 1$ for $i = k^*$. Therefore, we get $\lim_{t\to\infty}\text{diag}\left(\frac{u}{u_{k^*}}\right)^{t+1} = \text{diag}(e_{k^*})$ with $e_i \in \mathbb{R}^n$ defined as a n dimensional vector with the ith element equal 1 and all the others equal zero. Together with (1.10), we further get,

$$\begin{aligned}
\lim_{t\to\infty} x_2(t+1) &= \lim_{t\to\infty} \frac{\text{diag}\left(\frac{u}{u_{k^*}}\right)^{t+1}x_2(0)}{\|\text{diag}\left(\frac{u}{u_{k^*}}\right)^{t+1}x_2(0)\|} \\
&= \frac{\lim_{t\to\infty}\text{diag}\left(\frac{u}{u_{k^*}}\right)^{t+1}x_2(0)}{\|\lim_{t\to\infty}\text{diag}\left(\frac{u}{u_{k^*}}\right)^{t+1}x_2(0)\|}
\end{aligned}$$

$$
\begin{aligned}
&= \frac{\mathrm{diag}(e_{k^*})x_2(0)}{\|\mathrm{diag}(e_{k^*})x_2(0)\|} \\
&= \frac{x_{2k^*}(0)e_{k^*}}{\|x_{2k^*}(0)e_{k^*}\|} \\
&= \frac{x_{2k^*}(0)}{|x_{2k^*}(0)|}e_{k^*} \\
&= \begin{cases} e_{k^*} & \text{when } x_{2k^*}(0) > 0 \\ -e_{k^*} & \text{when } x_{2k^*}(0) < 0 \end{cases}
\end{aligned}
\tag{1.11}
$$

which means that $x_{2i}(t)$ converges to 0 for the losers with $x_{2i}(0) < x_{2k^*}$ and converges to 1 for the winner $i = k^*$. Recalling that $y(t) = x_2(t)$, the proof is completed.

1.5 Illustrative Examples

In this section, numerical examples are used to further explore the winner-take-all competition phenomena generated by the discrete-time dynamic (1.1). We consider two sceneries: one is static competition, where the input u is constant and one is dynamic competition, where the input u is time-varying.

1.5.1 Discrete-Time Static Competition

For the static competition problem, we consider a problem with $n = 10$ agents under time invariant input. The input u is randomly generated between 0 and 1, which is $u = [0.1982, 0.1951, 0.3268, 0.8803, 0.4711, 0.4040, 0.1792, 0.9689, 0.4075, 0.8445]$, and the state is randomly initialized between -1 and 1, Fig. 1.1 shows the evolution of output values of all agents with time (in the figure, the value of outputs is marked as '+' in each time step). From the figure, it can be observed that only a single output (corresponds to the 7th agent, which has the largest value in u) reaches 1 eventually and all the other output values are suppressed to zero. Figure 1.2 shows the evolution of output values of all agents with time with the same u but different initialization of states from that used in Fig. 1.1. In Fig. 1.2, all output values converges to zero except the output of the 7th agent, which has the largest value in u. It is noteworthy that the ultimate output value of the winner is 1 in Fig. 1.1 but is -1 in Fig. 1.2. This observation is consistent with the theoretical conclusion drawn in Theorem 1.3 since the 7th agent, which is the winner in this set of simulations, is initialized in x_2 with a positive value in Fig. 1.1 but a negative value in Fig. 1.2 as can be observed in the two figures.

We next consider a three-agent competition problem for the convenience of visualization. A three-agent system with $u = [0.5598, 0.3008, 0.9394]$ is simulated (in this case, the third agent has the largest input and thus is the winner). Figure 1.3

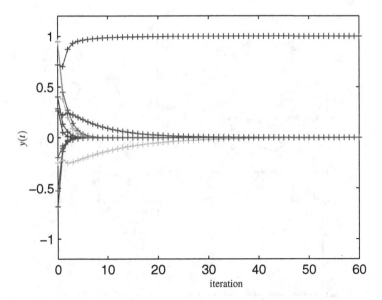

Fig. 1.1 The output of $y(t)$ in all dimensions in the static competition scenario under a random initialization with 10 agents

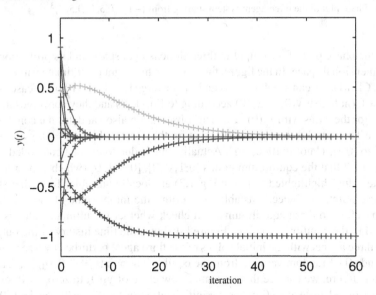

Fig. 1.2 The output of $y(t)$ in all dimensions in the static competition scenario with 10 agents under a different random initialization from Fig. 1.1

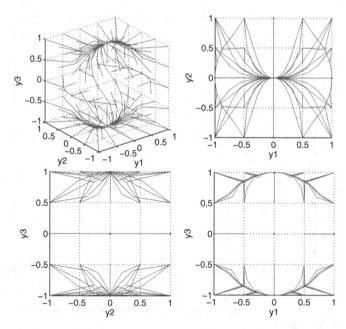

Fig. 1.3 Phase plot of the three-agent system starting from $\{-1, -0.5, 0, 0.5, 1\}^3$

shows the phase plot of the output in three-dimensional space and its projections in
two-dimensional space. In the figure, the red spots highlight the ultimate value of the
output. Clearly, we can see that y_1, y_2 and y_3 converge to 0, 0 and 1 for the cases with
$y_3(0) > 0$ (note that $y(0) = x_2(0)$ according to Eq. (1.2)) and they converges to 0, 0
and -1 for the cases with $y_3(0) < 0$. This observation also validates the conclusion
drawn in Theorem 1.3. It is noteworthy that for the cases with $y_3(0) = 0$, y converge
neither to $[0, 0, 1]$ nor to $[0, 0, -1]$. Actually, this is due to the fact concluded from
Theorem 1.2 that the equilibrium points $[\pm1, 0, 0]$, $[0, \pm1, 0]$ (which appear as the
ultimate values highlighted in red in Fig. 1.3) are locally unstable. To further show
that these points are indeed unstable, we perturb the initial value of the system to
one very close to these equilibriums, and check whether the ultimate value is still
attracted to the equilibrium points. Figure 1.4 shows the time history of the outputs
for the three agents with the initial values of the third agent perturbed by a zero mean
magnitude 0.01 random variable from the equilibrium points $[\pm1, 0, 0]$, $[0, \pm1, 0]$.
From this figure, we can see that no matter how close of $y_3(0)$ to zero, $y_3(0)$ either
goes to 1 or -1 instead of 0 under the initialization of $[y_1(0), y_2(0)] = [\pm1, 0]$ or
$[y_1(0), y_2(0)] = [0, \pm1]$ only if $y_3(0) \neq 0$.

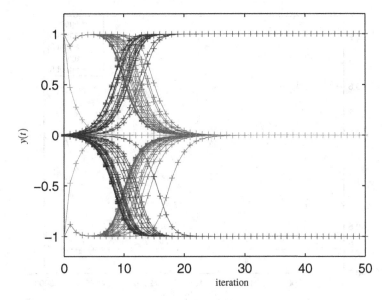

Fig. 1.4 Time history of the three-agent system with small perturbations on $y_3(0)$ around $[\pm1, 0, 0]$ or $[0, \pm1, 0]$

1.5.2 Discrete-Time Dynamic Competition

In this part, we consider the scenario with time-varying inputs. Note that the winner-take-all system should run in a greater sampling rate in order to successfully track time-varying signals. In the simulation, we consider $n = 5$ agents with input $u_i(t) = 2.5 + \sin\left(\frac{2\pi}{1000}t + \frac{2\pi}{5}i\right)$ for $i = 1, 2, 3, 4, 5$ and for $t = 0, 1, 2, \ldots$. The initial value of y in all dimensions are randomly generated as a positive number between 0 and 1 to guarantee the winner converges to 1 instead of -1. To avoid the output is stuck at the unstable equilibrium points due the computation error, we add an extra zero mean 0.005 magnitude random disturbance in the update of x_2 (the second equation in (1.1)). The five input signals and output $y(t)$ are plotted in Fig. 1.5. The figure implies the system can successfully find the winner in real time.

1.6 Summary

In this chapter, a simple dynamic system described by a difference equation are presented to reach the winner-take-all competition among agents. This model is described by a differential equation with continuous time dynamics instead of a difference equation with completely different discrete-time dynamics. The equilibrium points are solved analytically and their local stability is investigated theoretically. In

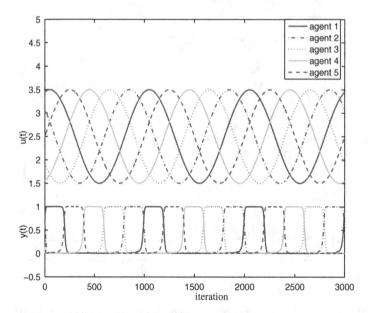

Fig. 1.5 Inputs and outputs of the dynamic system in the dynamic competition scenario

addition, global stability of the model is also rigorously studied in theory. Numerical simulations are performed and the results validate the effectiveness of the dynamic equation in describing the winner-take-all competition.

References

1. Dun EA, Ferguson JD, Beveridge CA (2006) Apical dominance and shoot branching. Divergent opinions or divergent mechanisms? Plant Physiol 142(3):812–819
2. Enquist M, Ghirlanda S (2005) Neural networks and animal behavior. Princeton University Press, Princeton
3. Jin L, Li S (2017) Distributed task allocation of multiple robots: A control perspective. IEEE Trans Syst Man Cybern Syst pp(99):1–9
4. Ramirez-Angulo J, Ducoudray-Acevedo G, Carvajal R, Lopez-Martin A (2005) Low-voltage high-performance voltage-mode and current-mode wta circuits based on flipped voltage followers. IEEE Trans Circuits Syst II Express Briefs 52(7):420–423
5. Emilio H, Lopez C, Pigolotti S, Andersen K (2008) Species competition: coexistence, exclusion and clustering. Philos Trans Roy Soc A Math Phys Eng Sci 367(3):3183–3195
6. Benkert C, Anderson DZ (1991) Controlled competitive dynamics in a photorefractive ring oscillator: Winner-takes-all and the voting-paradox dynamics. Phys Rev A 44(1):4633–4638
7. Jin L, Zhang Y, Li S, Zhang Y (2016) Modified ZNN for time-varying quadratic programming with inherent tolerance to noises and its application to kinematic redundancy resolution of robot manipulators. IEEE Trans Ind Electron 63(11):6978–6988
8. Jin L, Zhang Y (2015) Discrete-time Zhang neural network for online time-varying nonlinear optimization with application to manipulator motion generation. IEEE Trans Neural Netw Learn Syst 27(6):1525–1531

9. Li S, Li Y, Wang Z (2013) A class of finite-time dual neural networks for solving quadratic programming problems and its k-winners-take-all application. Neural Netw 39(1):27–39
10. Jin L, Zhang Y, Qiao T, Tan M, Zhang Y (2016) Tracking control of modified Lorenz nonlinear system using ZG neural dynamics with additive input or mixed inputs. Neurocomputing 196(1):82–94
11. Li S, Zhang Y, Jin L (2016) Kinematic control of redundant manipulators using neural networks. IEEE Trans Neural Netw Learn Syst. doi:10.1109/TNNLS.2016.2574363 (In Press)
12. Li S, Chen S, Liu B (2013) Accelerating a recurrent neural network to finite-time convergence for solving time-varying Sylvester equation by using a sign-bi-power activation function. Neurocomputing 37(1):189–205
13. Zhang Y, Yi C (2011) Zhang neural networks and neural-dynamic method. Nova Science Publishers, New York
14. Li S, Liu B, Li Y (2013) Selective positive-negative feedback produces the winner-take-all competition in recurrent neural networks. IEEE Trans Neural Netw Learn Syst 24(2):301–309
15. Li S, He J, Rafique U, Li Y (2017) Distributed recurrent neural networks for cooperative control of manipulators: A game-theoretic perspective. IEEE Trans Neural Netw Learn Syst 28(2):415–426
16. Jin L, Zhang Y, Li S (2016) Integration-enhanced Zhang neural network for real-time varying matrix inversion in the presence of various kinds of noises. IEEE Trans Neural Netw Learn Syst 27(12):2615–2627
17. Jin L, Zhang Y, Li S, Zhang Y (2017) Noise-tolerant ZNN models for solving time-varying zero-finding problems: A control-theoretic approach. IEEE Trans Autom Control 62(2):577–589
18. Jin L, Zhang Y (2016) Continuous and discrete Zhang dynamics for real-time varying nonlinear optimization. Numer Algorithm 73(1):115–140
19. Li S, Li Y (2014) Nonlinearly activated neural network for solving time-varying complex sylvester equation. IEEE Trans Cybern 44(8):1397–1407
20. Li S, You Z, Guo H, Luo X, Zhao Z (2016) Inverse-free extreme learning machine with optimal information updating. IEEE Trans Cybern 46(5):1229–1241
21. Khan M, Li S, Wang Q, Shao Z (2016) CPS oriented control design for networked surveillance robots with multiple physical constraints. IEEE Trans Comput-Aided Des Integr Circuits Syst 35(5):778–791
22. Khan M, Li S, Wang Q, Shao Z (2016) Formation control and tracking for co-operative robots with non-holonomic constraints. J Intell Rob Syst 82(1):163–174
23. Fangi Y, Cohen M, Kincaid T (2010) Dynamic analysis of a general class of winner-take-all competitive neural networks. IEEE Trans Neural Netw 21(5):771–783
24. Sum JPF, Cohen CS, Tam PKS, Young GH, Kan WK, Chan LW (1999) Analysis for a class of winner-take-all model. IEEE Trans Neural Netw 10(1):64–71
25. Wang W, Slotine J (2006) Fast computation with neural oscillators. Neurocomputing 69(1):2320–2326
26. Oster M, Douglas R, Liu S (2009) Computation with spikes in a winner-take-all network. Neural Comput 21(1):2437–2465
27. Xu Z, Jin H, Leung K, Wong CK (2002) An automata network for performing combinatorial optimization. Neurocomputing 47(1–4):59–83
28. Liu S, Wang J (2006) A simplified dual neural network for quadratic programming with its kwta application. IEEE Trans Neural Netw 17(6):1500–1510
29. Jin L, Li S, La H, Luo X (2017) Manipulability optimization of redundant manipulators using dynamic neural networks. IEEE Trans Ind Electron pp(99):1–10. doi:10.1109/TIE.2017.2674624 (In press)
30. Zhang Y, Li S (2017) Predictive suboptimal consensus of multiagent systems with nonlinear dynamics. IEEE Trans Syst Man Cybern Syst pp(99):1–11. doi:10.1109/TSMC.2017.2668440 (In press)
31. Jin L, Zhang Y, Qiu B (2016) Neural network-based discrete-time Z-type model of high accuracy in noisy environments for solving dynamic system of linear equations. Neural Comput Appl. doi:10.1007/s00521-016-2640-x (In press)

32. Li S, Zhou M, Luo X, You Z (2017) Distributed winner-take-all in dynamic networks. IEEE Trans Autom Control 62(2):577–589
33. Jin L, Zhang Y (2015) G2-type SRMPC scheme for synchronous manipulation of two redundant robot arms. IEEE Trans Cybern 45(2):153–164
34. Li S, Cui H, Li Y (2013) Decentralized control of collaborative redundant manipulators with partial command coverage via locally connected recurrent neural networks. Neural Comput Appl 23(1):1051–1060
35. Clark L, Cools R, Robbins TW (2004) The neuropsychology of ventral prefrontal cortex: Decision-making and reversal learning. Brain Cogn 55(1):41–53

Chapter 2
Competition Aided with Continuous-Time Nonlinear Model

Abstract In this chapter, different from the model presented in Chap. 1, we present a continuous-time dynamic model, which is described by an ordinary differential equation and is able to produce the winner-take-all competition by taking advantage of selective positive-negative feedback. The global convergence is proven analytically and the convergence rate is also discussed. Simulations are conducted in the static competition and the dynamic competition scenarios. Both theoretical and numerical results validate the effectiveness of the dynamic equation in describing the nonlinear phenomena of winner-take-all competition.

Keywords Winner-take-all competition · Recurrent neural networks · Continuous-time system · Global stability · Selective positive-negative feedback

2.1 Introduction

Competition widely exists in nature and the society. Among different kinds of competitions, winner-take-all competition refers to the phenomena that individuals in a group compete with each others for activation and only the one with the highest input stays activated while all the others deactivated. Examples of this type of competition include the dominant growth of the central stem over others [1], the contrast gain in the visual systems through a winner-take-all competition among neurons [2], competitive decision making in the cortex [3], competition-based coordination control of robots [4], etc.

Although many phenomena, as exemplified above, demonstrate the same winner-take-all competition, they may have different underlying principles in charge of the dynamic evolution. Apart from the natures of distributed-storage and high-speed parallel-processing, neural networks can be readily implemented by hardware and thus have been widely applied in various fields, including the competition phenomena [5–25]. For example, the N species Lotka-Volterra model [26], interactively spiking FitzHugh-Nagumo Model [27], discrete-time different equation model presented in Chap. 1. However, these models are often very complicated due to the compromise with experimental realities in the particular fields. Consequently, the essence of the winner-take-all competition may be embedded in the interaction dynamics of those

S. Li and L. Jin, *Competition-Based Neural Networks with Robotic Applications*,
SpringerBriefs in Applied Sciences and Technology,
DOI 10.1007/978-981-10-4947-7_2

models, but difficult to tell from the sophisticated dynamic equations. Motivated by this, a simple ordinary differential equation model with a direct and intuitive explanation is presented in this chapter and it is expected to cast lights to researchers on the principle of competition phenomena in different fields.

2.2 The Model

The presented model has the following dynamic for the ith agent in a group of totally n agents,

$$\dot{x}_i = c_0(u_i - \|x\|^2)x_i \tag{2.1}$$

where $x_i \in \mathbb{R}$ denotes the state of the i agent, $u_i \in \mathbb{R}$ is the input and $u_i \geq 0$, $u_i \neq u_j$ for $i \neq j$, $\|x\| = \sqrt{x_1^2 + x_2^2 + \ldots + x_n^2}$ denotes the Euclidean norm of the state vector $x = [x_1, x_2, \ldots, x_n]^T$, $c_0 \in \mathbb{R}$ $c_0 \geq 0$ is a scaling factor.

The dynamic equation (2.1) can be written into the following compact form by stacking up the state for all agents,

$$\dot{x} = c_0(u \circ x - \|x\|^2 x) \tag{2.2}$$

where $x = [x_1, x_2, \ldots, x_n]^T$, $u = [u_1, u_2, \ldots, u_n]^T$, the operator '$\circ$' represents the multiplication in component-wise, i.e., $u \circ x = [u_1 x_1, u_2 x_1, \ldots, u_n x_n]^T$.

Remark 2.1 In the dynamic equation (2.1), all quantities on the right hand side can be obtained locally from the ith agent itself (u_i and x_i) except the quantity $\|x\|^2$, which reflects the effort from other agents over the ith one (as sketched in Fig. 2.1).

Fig. 2.1 Information flow for the agent dynamics

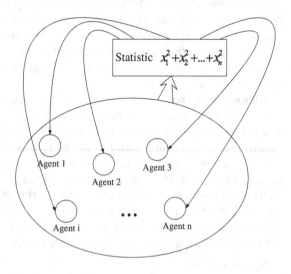

Actually, $\|x\|^2 = x_1^2 + x_2^2 + \ldots + x_n^2$ is the second moment about the origin of the group of agents and it is a statistic of the whole group. In this regard, the dynamic model (2.1) implies that the winner-take-all competition between agents may emerge in a multi-agent system if each agent accesses the global statistic $\|x\|^2$ (instead of exactly knowing states of all the other agents) besides its own information.

As will be stringently demonstrated later, the agent with the largest input will finally win the competition and keep active while all the other agents will be deactivated to zero eventually. Before proving this result rigorously, we present a intuitive explanation of the result in a sense of positive feedback versus negative feedback. Note that the term $c_0 u_i x_i$ in Eq. (2.1) provides a positive feedback to the state variable x_i as $u_i \geq 0$ while the term $-c_0\|x\|^2 x_i$ supplies a negative feedback. For the ith agent, if $u_i = \|x\|^2$, x_i will keep the value. If $u_i < \|x\|^2$, the positive feedback is less than the negative feedback in value and the state value attenuates to zero. In contrast, if $u_i > \|x\|^2$, the positive feedback is greater than the negative feedback and the state value tends to increase as large as possible until the resulting increase of $\|x_i\|$ surpasses u_i. Particularly for the winner, say the k^*th agent, $u_{k^*} > u_i$ holds for all $i \neq k^*$. In this case, all agents have negative feedbacks and keep reducing in values until $\|x\|^2$ reduces to the value of u_k when $u_k < \|x\|^2$. Otherwise when u_k is slightly greater than $\|x\|^2$ (by slightly greater we mean $u_k > \|x\|^2 > u_l$ with l denoting the agent with the second largest state value), only the winner has a positive feedback and has an increase in its state value while all the other agents have negative feedbacks and keep reducing until $\|x\|^2$ equals u_k. Under this selective positive-negative feedback mechanism, the winner finally stays active at the value $u_{k^*} = \|x\|^2$ while the losers are deactivated to zero.

2.3 Theoretical Analysis and Results

In this section, theoretical results on the dynamic system (2.1) are presented. The rigorous proof of the main results needs the uses of LaSalle's invariant set principle [28, 29], local stability analysis and the ultimate boundedness theory [30].

Lemma 2.1 ([28]) *Let* $\mathbb{D} \subset \mathbb{R}^n$ *be a domain that contains the origin and* $V :$ $[0, \infty) \times \mathbb{D} \to \mathbb{R}$ *be a continuous differentiable function such that*

$$\alpha_1(\|x\|) \leq V(t, x) \leq \alpha_2(\|x\|) \tag{2.3}$$

$$\dot{V} = \frac{\partial V}{\partial t} + \frac{\partial V}{\partial x} f(t, x) \leq -W(x), \quad \forall \|x\| \geq \mu > 0 \tag{2.4}$$

$\forall t \geq 0$ *and* $\forall x \in \mathbb{D}$, *where* α_1 *and* α_2 *are class* $\mathscr{K}$ *functions and* $W(x)$ *is a continuous positive definite function. Take* $r > 0$ *such that* $\mathbb{B}_r \subset \mathbb{D}$ *and suppose that* $\mu < \alpha_2^{-1}(\alpha_1(r))$. *Then, there exists a class* $\mathscr{K}\mathscr{L}$ *function* β *and for every initial state* $x(t_0)$, *satisfying* $\|x(t_0)\| \leq \alpha_2^{-1}(\alpha_1(r))$, *there is* $T \geq 0$ *(dependent on* $x(t_0)$ *and* μ) *such that the solution of* $\dot{x} = f(t, x)$ *satisfies,*

$$\|x(t)\| \leq \alpha_1^{-1}(\alpha_2(\mu)) \quad \forall t \geq t_0 + T \tag{2.5}$$

Moreover, if $\mathbb{D} = \mathbb{R}^n$ *and* α_1 *belongs to class* $\mathcal{K}_\infty$, *then the result holds for any initial state* $x(t_0)$, *with no restriction on how large* μ *is.*

With Lemma 2.1, we are able to prove the following lemma for our main result,

Lemma 2.2 *There exists* $T \geq 0$ *(dependent on* $x(t_0)$ *and* μ*) such that the solution of the agent dynamic equation (2.2) satisfies,*

$$\|x(t)\| \leq \mu \quad \forall t \geq t_0 + T \tag{2.6}$$

where $\mu = \sqrt{\max\{u_1, u_2, ..., u_n\}} + \delta$ *with* $\delta > 0$ *being any positive constant.*

Proof We prove the result by following the framework of Lemma 2.1. Let $\mathbb{D} = \mathbb{R}^n$, $V = \frac{1}{2} x^T x$ and $\alpha_1(\|x\|) = \alpha_2(\|x\|) = \frac{1}{2}\|x\|^2 = V$. For V, we have,

$$\begin{aligned}
\dot{V} &= x^T \dot{x} \\
&= c_0 x^T (u \circ x - \|x\|^2 x) \\
&= c_0 x^T (diag(u)x - \|x\|^2 x) \\
&= c_0 x^T (diag(u) - \|x\|^2)x \\
&\leq c_0 (u_0 - \|x\|^2) x^T x \tag{2.7}
\end{aligned}$$

The equation $u \circ x = \mathrm{diag}(u)x$ is used in the second step of the above derivation. Note that $(diag(u) - \|x\|^2)$ is a diagonal matrix and its largest eigenvalue is $u_0 - \|x\|^2$. Therefore, $x^T (diag(u) - \|x\|^2)x \leq (u_0 - \|x\|^2)x^T x$, from which the last inequality in (2.7) is obtained. As $u_i \geq 0$ for all i and $u_i \neq u_j$ for $i \neq j$, we get $u_0 > 0$. Recall $\mu = \sqrt{\max\{u_1, u_2, ..., u_n\}} + \delta$, i.e., $\mu = \sqrt{u_0} + \delta$ for any small positive $\delta > 0$. For $\|x\| \geq \mu$, $u_0 - \|x\|^2 \leq -\delta^2$. Together with (2.7), we get,

$$\dot{V} \leq -c_0 \delta^2 x^T x \tag{2.8}$$

for $\|x\| \geq \mu$. Choosing a positive definite function $W(x) = c_0 \delta^2 x^T x$ yields $\dot{V} \leq -W(x)$ for $\forall \|x\| \geq \mu$. Therefore, according to Lemma 2.1, there exists $T \geq 0$ such that the solution satisfies $\|x(t)\| \leq \alpha_1^{-1}(\alpha_2(\mu)) = \mu$, $\forall t \geq t_0 + T$. This completes the proof.

Remark 2.2 Lemma 2.2 means the state of the dynamic model (2.2) is ultimately bounded inside a compact super ball in $\mathbb{R}^n$ with radius $\mu = \sqrt{\max\{u_1, u_2, ..., u_n\}} + \delta$. In other words, this super ball is positively invariant with respect the system dynamic (2.2). This result allows us to apply LaSalle's invariant set principle for further investigation of the system behaviors.

Lemma 2.3 ([28]) *Let* $\Omega \subset \mathbb{D}$ *be a compact set that is positively invariant with respect to* $\dot{x} = f(x)$. *Let* $V : \mathbb{D} \to \mathbb{R}$ *be a* C^1-*function such that* $\dot{V}(x) \leq 0$ *on* Ω.

Let $\mathbb{E}$ *be the set of all points in* Ω *such that* $\dot{V}(x) = 0$. *Let* $\mathbb{M}$ *be the largest invariant set in* $\mathbb{E}$. *Then, every solution starting in* Ω *approaches* $\mathbb{M}$ *as* $t \to \infty$.

Remark 2.3 It is worth noting that the mapping V in Lemma 2.3 is not necessary to be positive definite, which is a major difference from the Lyapunov function in conventional stability analysis of dynamic systems [28]. Instead, V is required to be be a continuous differentiable function in Lemma 2.3, which is much looser than the positive definite requirement and simplifies the analysis.

Theorem 2.1 *The solution of the system involving* n *dynamic agents with the* i*th agent described by* (2.1) *globally approaches* 0 *for* $i \neq k^*$ *and approaches* $\sqrt{u_{k^*}}$ *or* $-\sqrt{u_{k^*}}$ *for* $i = k^*$ *as* $t \to \infty$, *where* k^* *denotes the label of the winner, i.e.,* $k^* = argmax\{u_1, u_2, ..., u_n\}$.

Proof There are two steps for the proof. The first step is to prove that the state variable ultimately converges to a set consisting of a limit number of points and the second step proves there is only a single point among the candidates is stable.

Step 1: According to previously presented Lemma 2.2, the state variable x in the system dynamic (2.2) is ultimately bounded by a compact super ball in $\mathbb{R}^n$ with radius $\mu = \sqrt{\max\{u_1, u_2, ..., u_n\}} + \delta$, which implies this super ball is positively invariant with respect the system dynamic (2.2) and the super ball $\{x \in \mathbb{R}^n | \|x\| \leq \mu\}$ is qualified to be the set Ω in Lemma 2.3.

Let $V = -\frac{1}{2}x^T \text{diag}(u)x + \frac{1}{4}\|x\|^4$. Apparently, V is a C^1-function. For V, we have,

$$\dot{V} = -x^T \text{diag}(u)\dot{x} + \|x\|^2 x^T \dot{x}$$
$$= \left(-x^T \text{diag}(u) + \|x\|^2 x^T\right)\dot{x} \tag{2.9}$$

With $x^T \text{diag}(u) = (x \circ u)^T$, we get $x^T \text{diag}(u) - \|x\|^2 x^T = (x \circ u - \|x\|^2 x)^T$. Together with (2.9), we have,

$$\dot{V} = -c_0(x \circ u - \|x\|^2 x)^T(x \circ u - \|x\|^2 x)$$
$$= -c_0 \|x \circ u - \|x\|^2 x\|^2$$
$$\leq 0 \tag{2.10}$$

We find $\text{diag}(u)x = \|x\|^2 x$ by letting $\dot{V} = 0$. Note that $\text{diag}(u)x = \|x\|^2 x$ is actually a eigenvector equation relative to the matrix $\text{diag}(u)$. The solution can be solved as the set $\mathbb{M} = \{0, \pm\sqrt{u_i}e_i$ for $i = 1, 2, ..., n\}$, where e_i is a n-dimensional vector with the ith component 1 and all the other component 0. According to Lemma 2.3, every solution starting in $\Omega = \{x \in \mathbb{R}^n | \|x\| \leq \mu\}$ approaches $\mathbb{M}$ as $t \to \infty$. Together with the fact proven in Lemma 2.2 that every solution stays in Ω ultimately, we conclude that every solution with the initialization $x(t_0) \in \mathbb{R}^n$ approaches $\mathbb{M}$ as $t \to \infty$.

Step 2: We have shown that there are several candidate fixed points to stay for the dynamic system. In this step, we show that all those fixed points in $\mathbb{M}$ are unstable except $x = \pm\sqrt{u_k}e_k$, where $k^* = argmax\{u_1, u_2, ..., u_n\}$. Lyapunov's indirect method suffices the analysis of the un-stability.

For the fixed point $x_e = 0$, the system dynamic (2.2) is linearized as $\dot{x} = c_0 \text{diag}(u)x$ about $x = 0$ and is unstable as the eigenvalues of the system matrix $c_0 \text{diag}(u)$ have positive real parts.

For the fixed points $x_e = \pm\sqrt{u_i}e_i$, the linearized system around the fixed point is as follows,

$$\dot{x} = c_0\left(\text{diag}(u) - 2x_e x_e^T - \|x_e\|^2\right)x \tag{2.11}$$

The system matrix of the above system is a diagonal matrix and its jth diagonal component, which is also its jth eigenvalue, is $c_0(u_j - u_i)$ for $j \neq i$ and $-2c_0$ for $j = i$. Clearly, all the eigenvalues have negative real part only when $u_j - u_i < 0$ holds for all $j \neq i$, i.e., when $i = k^*$, which excludes all fixed points except for $x_e = \pm\sqrt{u_{k^*}}e_{k^*}$ from the stable ones.

In summary, we conclude that every solution approaches $x = \pm\sqrt{u_{k^*}}e_{k^*}$ ultimately with $k^* = \text{argmax}\{u_1, u_2, ..., u_n\}$ and e_{k^*} being a n-dimensional vector with the k^*th component 1 and all the other component 0. Entrywisely, the solution approaches $x_i = 0$ for $i \neq k^*$ and $x_{k^*} = \pm\sqrt{u_{k^*}}$, which completes the proof.

Remark 2.4 According to Theorem 2.1, the steady-state value of the winner is either $\sqrt{u_{k^*}}$ or $-\sqrt{u_{k^*}}$. Actually, we can conclude that it is $\sqrt{u_{k^*}}$ if the initial state of the winner is positive while it is $-\sqrt{u_{k^*}}$ if the initial value is negative by noting that $\dot{x}_{k^*} = 0$ when $x_{k^*} = 0$ in (2.1) for $i = k^*$, which means the state value x_{k^*} will never cross the critical value $x^* = 0$.

2.4 Illustrative Examples

In this section, simulations are provided to illustrate the the winner-take-all competition phenomena generated by the agent dynamic (2.1). We consider two sceneries: one is static competition, i.e., the input u is constant and one is dynamic competition, i.e., the input u is time-varying.

2.4.1 Static Competition

For the static competition problem, we consider time invariant signals as the input. In the simulation, we consider a problem with $n = 15$ agents. The input u is randomly generated between 0 and 1, which is $u =$ [0.0924, 0.0078, 0.4231, 0.6556, 0.7229, 0.5312, 0.1088, 0.6318, 0.1265, 0.1343, 0.0986, 0.1420, 0.1683, 0.1962, 0.3175], and the state is randomly initialized between -1 and 1, which is $x(0) =$ [0.7556, 0.1649, -0.8586, 0.8455, 0.6007, -0.4281, 0.0873, 0.9696, 0.4314, 0.6779, -0.1335, -0.0588, 0.1214, -0.4618, 0.4980]. In the simulation, we choose the scaling factor $c_0 = 1$. Figure 2.2 shows the evolution of state values of all agents with time, from which it can be observed that only a single state (cor-

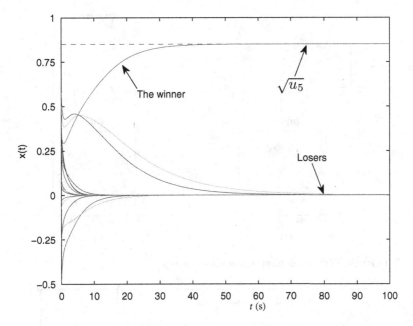

Fig. 2.2 Agent state trajectories in the static competition scenario with 15 agents

responds to the 5th agent, which has the largest value in u) has a non-zero value eventually and all the other state values are suppressed to zero. Also, the value of x_5 approaches $\sqrt{u_5}$ (see Fig. 2.2), which is consistent with the claim made in Remark 2.4 since the initial value $x_5(0) = 0.6007 > 0$.

To fully visualize the interaction between agents, we consider a three agent case with $u = [0.7368, 0.2530, 0.4117]$. Figure 2.3 shows the phase plot of the state in three-dimensional space and its projections in two-dimensional space. Clearly, we can see that the states with the initial state value of the winner being negative (i.e., $x_1(0) < 0$) is attracted to $[-\sqrt{u_1}, 0, 0]$ while is attracted to $[\sqrt{u_1}, 0, 0]$ for the cases with positive initial state values of the winner (i.e., $x_1(0) > 0$). It is worth noting that $x_1(t)$ appears staying at 0 in the situation with $x_1(0) = 0$ in Fig. 2.3, which seems in contradiction with the statement that the winner $x_1(t)$ converges to either $\sqrt{u_1}$ or $-\sqrt{u_1}$ eventually. Actually, as mentioned in the proof of the Theorem 2.1, all fixed points are unstable except $\pm\sqrt{u_{k^*}}e_{k^*}$ (k^* denotes the label of the winner and e_{k^*} is a n dimensional vector with the k^*th element being 1 and all the other elements being zeros). Therefore, in this case, the state with $x_1 = 0$ is unstable and must be very subjective to disturbances. To show this, we plug a small random Gaussian white noise with zero mean and 0.0001 variance into the agent dynamic (2.1). In equation, the resulting dynamic for the ith agent is $\dot{x}_i = c_0(u_i - \|x\|^2)x_i + 0.0001v_i$, where v_i is a Gaussian white noise with zero mean unit variance and it is independent with v_j for $j \neq i$. Even with such a small perturbation with a magnitude of 0.0001,

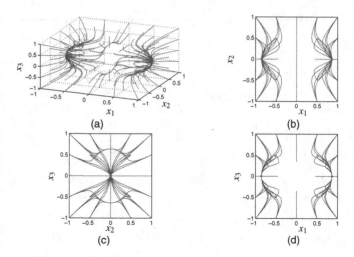

Fig. 2.3 Phase plot of the three-agent system without noises

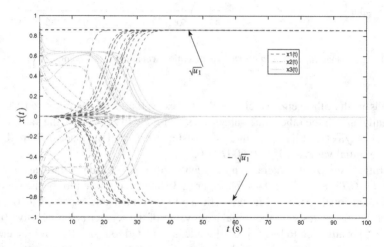

Fig. 2.4 Time history of the three-agent system with small perturbations

the states with $x_1(0) = 0$ either converge to $[\sqrt{u_1}, 0, 0]$ or $[-\sqrt{u_1}, 0, 0]$ instead of staying at $x_1 = 0$ as shown in Figs. 2.4 and 2.5, where the state is initialized at $x(t_0) = 0 \times [-1, -0.5, 0, 0.5, 1]^2$.

2.4.2 Dynamic Competition

In this part, we consider the scenario with time-varying inputs. For the dynamic system (2.1), the convergence can be accelerated by choosing a large scaling factor c_0, and the resulting fast response allows the computation of x in real time with

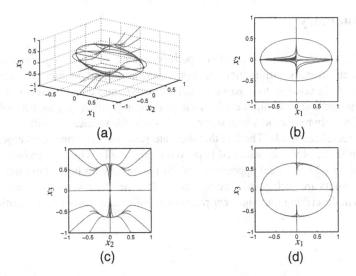

(a)

(b)

(c)

(d)

Fig. 2.5 Phase plot of the three-agent system with small perturbations

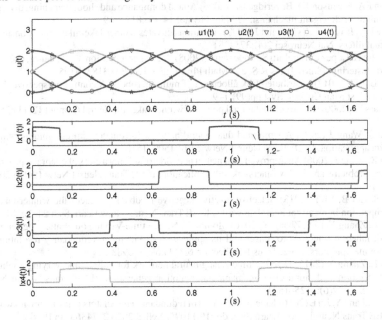

Fig. 2.6 Inputs and outputs of the dynamic system in the dynamic competition scenario

time-varying input $u(t)$. In this simulation, we choose $c_0 = 10^4$ and consider $n = 4$ agents with input $u_i(t) = 1 + \sin(2\pi t + 0.25i)$ for $i = 1, 2, 3, 4$, respectively. The initial state valued are randomly generated between -1 and 1. The four input signals and the absolute value of the state variables are plotted in Fig. 2.6. From this figure, we can see the system can successfully find the winner in real time.

2.5 Summary

In this chapter, the winner-take-all competition among agents in a group is considered and an ordinary differential equation describing the dynamics of each agent is presented. In contrast to existing models, this dynamic equation features a simple expression and an explicit explanation of the competition mechanism, which is expected to help researchers gain some insights into the winner-take-all phenomena in their specialized fields. The fact that the state value of the winner converges to be active while the others deactivated is proven theoretically. The convergence rate is discussed based on a local approximation. Simulations with both static inputs and dynamic inputs are performed. The results validate the effectiveness of the dynamic equation in describing the nonlinear phenomena of winner-take-all competition.

References

1. Dun EA, Ferguson JD, Beveridge CA (2006) Apical dominance and shoot branching. divergent opinions or divergent mechanisms? Plant Physiol 142(3):812–819
2. Lee DK, Itti L, Koch C, Braun J (1999) Attention activates winner-take-all competition among visual filters. Nat Neurosci 2(4):375–381
3. Emilio H, Lopez C, Pigolotti S, Andersen K (2008) Species competition: coexistence, exclusion and clustering. Philos Trans R Soc A Math Phys Eng Sci 367(3):3183–3195
4. Jin L, Li S (2017) Distributed task allocation of multiple robots: A control perspective. IEEE Trans Syst Man Cybern: Syst pp(99):1–9
5. Maass W (2000) On the computational power of winner-take-all. Neural Comput 12(11):2519–2535
6. Liu S, Wang J (2006) A simplified dual neural network for quadratic programming with its kwta application. IEEE Trans Neural Netw 17(6):1500–1510
7. Hu X, Wang J (2006) An improved dual neural network for solving a class of quadratic programming problems and its k-winners-take-all application. IEEE Trans Neural Netw 19(12):2022–2031
8. Li S, Liu B, Li Y (2013) Selective positive-negative feedback produces the winner-take-all competition in recurrent neural networks. IEEE Trans Neural Netw Learn Syst 24(2):301–309
9. Jin L, Zhang Y, Li S, Zhang Y (2016) Modified ZNN for time-varying quadratic programming with inherent tolerance to noises and its application to kinematic redundancy resolution of robot manipulators. IEEE Trans Ind Electron 63(11):6978–6988
10. Jin L, Zhang Y (2015) Discrete-time Zhang neural network for online time-varying nonlinear optimization with application to manipulator motion generation. IEEE Trans Neural Netw Learn Syst 27(6):1525–1531
11. Li S, Zhang Y, Jin L (2016) Kinematic control of redundant manipulators using neural networks. IEEE Trans Neural Netw Learn Syst. doi:10.1109/TNNLS.2016.2574363 (In Press)
12. Li S, He J, Rafique U, Li Y (2017) Distributed recurrent neural networks for cooperative control of manipulators: a game-theoretic perspective. IEEE Trans Neural Netw Learn Syst 28(2):415–426
13. Jin L, Zhang Y, Li S (2016) Integration-enhanced Zhang neural network for real-time varying matrix inversion in the presence of various kinds of noises. IEEE Trans Neural Netw Learn Syst 27(12):2615–2627
14. Jin L, Zhang Y, Li S, Zhang Y (2017) Noise-tolerant ZNN models for solving time-varying zero-finding problems: a control-theoretic approach. IEEE Trans Autom Control 62(2):577–589

15. Jin L, Zhang Y (2016) Continuous and discrete Zhang dynamics for real-time varying nonlinear optimization. Numer Algorithms 73(1):115–140

16. Li S, Li Y (2014) Nonlinearly activated neural network for solving time-varying complex sylvester equation. IEEE Trans Cybern 44(8):1397–1407

17. Jin L, Li S, La H, Luo X (2017) Manipulability optimization of redundant manipulators using dynamic neural networks. IEEE Trans Ind Electron pp(99):1–10 In press with doi:10.1109/TIE.2017.2674624

18. Zhang Y, Li S (2017) Predictive suboptimal consensus of multiagent systems with nonlinear dynamics. IEEE Trans Syst Man Cybern: Syst pp(99):1–11, In press with doi:10.1109/TSMC.2017.2668440

19. Jin L, Zhang Y, Qiu B (2016) Neural network-based discrete-time Z-type model of high accuracy in noisy environments for solving dynamic system of linear equations. Neural Comput Appl. In press with doi:10.1007/s00521-016-2640-x

20. Li S, Zhou M, Luo X, You Z (2017) Distributed winner-take-all in dynamic networks. IEEE Trans Autom Control 62(2):577–589

21. Jin L, Zhang Y (2015) G2-type SRMPC scheme for synchronous manipulation of two redundant robot arms. IEEE Trans Cybern 45(2):153–164

22. Li S, Cui H, Li Y (2013) Decentralized control of collaborative redundant manipulators with partial command coverage via locally connected recurrent neural networks. Neural Comput Appl 23(1):1051–1060

23. Li S, You Z, Guo H, Luo X, Zhao Z (2016) Inverse-free extreme learning machine with optimal information updating. IEEE Trans Cybern 46(5):1229–1241

24. Khan M, Li S, Wang Q, Shao Z (2016) CPS oriented control design for networked surveillance robots with multiple physical constraints. IEEE Trans Computer-Aided Des Integr Circuits Syst 35(5):778–791

25. Khan M, Li S, Wang Q, Shao Z (2016) Formation control and tracking for co-operative robots with non-holonomic constraints. J Intell Robot Sys 82(1):163–174

26. Ramirez-Angulo J, Ducoudray-Acevedo G, Carvajal R, Lopez-Martin A (2005) Low-voltage high-performance voltage-mode and current-mode wta circuits based on flipped voltage followers. IEEE Trans Circuits Syst II: Express Br 52(7):420–423

27. Benkert C, Anderson DZ (1991) Controlled competitive dynamics in a photorefractive ring oscillator: winner-takes-all and the voting-paradox dynamics. Phys Rev A 44(1):4633–4638

28. Isidori A (1995) Nonlinear control systems. Springer, New York

29. LaSalle JP, Lefschetz S (1973) Stability by Liapunov's direct method with applications. Academic Press, New York

30. Khalil H (2002) Nonlinear systems. Prentice Hall, New York

Chapter 3
Competition Aided with Finite-Time Neural Network

Abstract In this chapter, a class of recurrent neural networks to solve quadratic programming problems are presented and further extended to competition generation. Different from most existing recurrent neural networks for solving quadratic programming problems, the proposed neural network model converges in finite time and the activation function is not required to be a hard-limiting function for finite convergence time. The stability, finite-time convergence property and the optimality of the proposed neural network for solving the original quadratic programming problem are proven in theory. Extensive simulations are performed to evaluate the performance of the neural network with different parameters. In addition, the proposed neural network is applied to solving the k-winner-take-all (k-WTA) problem. Both theoretical analysis and numerical simulations validate the effectiveness of our method for solving the k-WTA problem.

Keywords Winner-take-all competition · Recurrent neural networks · Quadratic programming · Global stability · Finite-time convergence · Numerical simulations

3.1 Introduction

In the past two decades, recurrent neural networks have received considerable studies in many scientific and engineering fields, such as motion planning of redundant robot manipulators [1], nonlinear optimization [2, 3], tracking control of chaotic systems [4], kinematic control of redundant manipulators [5, 6], etc. Particularly, after the invention of the well-known Hopfield neural network, which was originally designed for real-time optimization, the recurrent neural network, as a powerful online optimization tool with potential parallel implementations, is becoming an independent research direction in online optimization field [7–12].

Remarkable advances have been made in the area of recurrent neural networks for online optimization [13–18]. To a constrained optimization problem, early works, such as [19], often remove the explicit constraints by introducing a penalty term into the cost function and then design a recurrent neural network evolving along the gradient descent direction. This type of neural networks only converges to an

© The Author(s) 2018

S. Li and L. Jin, *Competition-Based Neural Networks with Robotic Applications*,
SpringerBriefs in Applied Sciences and Technology,
DOI 10.1007/978-981-10-4947-7_3

approximation of the optimal solution. In order to obtain a recurrent neural network with guaranteed convergence to the optimal solution, later works, such as [20, 21], introduce dynamic Lagrange multipliers to regulate the constraints. Compared to early works using penalty strategies, both [20, 21] are able to converge to the optimal solution of the constrained optimization problem, but the number of neurons in the neural network is increased since extra neurons are required for the dynamics of the Lagrange multipliers. Because the number of neurons in the neural network is directly relevant to the complexity and cost of its hardware implementation, some researchers turned their attention to the reduction of neuron number in the design. Typical works include [22–25], which consider the problem in the dual space and use a projection function to represent inequality constraints. This type of method significantly simplifies the architecture without losing efficiency or accuracy and is successfully used in various applications, such as kinematic control of redundant manipulators [26], k-winner-take-all (k-WTA) problem solving [23], etc. Although most of the above mentioned neural networks are stable or even with an exponential convergence rate, they never converge in finite time. Realizing this point, some neural networks with finite-time convergence properties are explored, for example, [27, 28], namely. To obtain the finite-time convergence, both [27, 28] use a discontinuous hard-limiting function as the activation function. Differently, inspired by the study on finite-time stability of autonomous systems [29], in this chapter we propose a class of recurrent neural networks, which uses a continuous activation function, but still has finite-time convergence to the optimal solution of the problem.

3.2 Model Description

In this chapter, we study the following quadratic programming problem:

$$\text{minimize} \quad \tfrac{1}{2}x^T W x + c^T x \tag{3.1a}$$

$$\text{subject to} \quad Ax = b \tag{3.1b}$$

$$l \leq Ex \leq h \tag{3.1c}$$

where $x \in \mathbb{R}^n$, $W \in \mathbb{R}^{n \times n}$ is a positive definite matrix, $c \in \mathbb{R}^n$, $A \in \mathbb{R}^{m \times n}$, $b \in \mathbb{R}^m$, $E \in \mathbb{R}^{q \times n}$, $h \in \mathbb{R}^q$, $l \in \mathbb{R}^q$, $m < n$ and $h \geq l$. Following the tradition [23], we assume that the equality constraint is irredundant, i.e., rank(A) = m.

According to Karash-Kuhn-Tucker (KKT) conditions [30], the solution to problem (3.1) satisfies,

$$Wx + c + A^T \lambda + E^T \mu = 0 \tag{3.2a}$$

$$Ax = b \tag{3.2b}$$

$$\begin{cases} Ex = h & \text{if } \mu > 0 \\ l \le Ex \le h & \text{if } \mu = 0 \\ Ex = l & \text{if } \mu < 0 \end{cases} \qquad (3.2c)$$

where $\lambda \in \mathbb{R}^m$ and $\mu \in \mathbb{R}^q$ are dual variables to the equality constraint (3.1b) and the inequality constraint (3.1c), respectively. By introducing a saturation function, (3.2c) can be simplified to

$$\rho Ex = g(\rho Ex + \mu) \qquad (3.3)$$

where $\rho \in \mathbb{R}, \rho > 0$ is a scaling factor, and the saturation function $g(x) = [g_1(x_1), g_2(x_2), \ldots, g_q(x_q)]^T$ is defined as

$$g_i(x_i) = \begin{cases} \rho h_i & \text{if } x_i > \rho h_i \\ x_i & \text{if } \rho l_i \le x_i \le \rho h_i \\ \rho l_i & \text{if } x_i < \rho l_i \end{cases} \qquad (3.4)$$

Recalling that W is positive definite and $\text{rank}(A) = m$, we can get the expression of x and λ explicitly in terms of μ by solving (3.2a) and (3.2b):

$$x = -\left(W^{-1}E^T - W^{-1}A^T(AW^{-1}A^T)^{-1}AW^{-1}E^T\right)\mu \\ - W^{-1}c + W^{-1}A^T(AW^{-1}A^T)^{-1}(b + AW^{-1}c) \qquad (3.5a)$$
$$\lambda = -(AW^{-1}A^T)^{-1}AW^{-1}E^T\mu - (AW^{-1}A^T)^{-1}(b + AW^{-1}c) \qquad (3.5b)$$

Note that $AW^{-1}A^T$ is invertible as $AW^{-1}A^T$ has full rank ($\text{rank}(AW^{-1}A^T) = \text{rank}(A) = m$). We define the following constant vector and matrix to simplify the expression of (3.5a),

$$s = W^{-1}A^T(AW^{-1}A^T)^{-1}(b + AW^{-1}c) - W^{-1}c$$
$$M = W^{-1} - W^{-1}A^T(AW^{-1}A^T)^{-1}AW^{-1} \qquad (3.6)$$

With this definition, x can be re-written as

$$x = -ME^T\mu + s \qquad (3.7)$$

Plugging (3.7) into (3.3), we have,

$$-\rho EME^T\mu + \rho Es = g\left((I - \rho EME^T)\mu + \rho Es\right) \qquad (3.8)$$

We use a layer of dynamic neurons to solve μ in (3.8) as follows

$$\varepsilon\dot{\mu} = -\text{sig}^r\left(g\left((I - \rho EME^T)\mu + \rho Es\right) + \rho EME^T\mu - \rho Es\right) \qquad (3.9)$$

where $\varepsilon \in \mathbb{R}$, $\varepsilon > 0$ is a scaling parameter, $r \in \mathbb{R}$, $0 < r < 1$. For $z \in \mathbb{R}^q$, $z = [z_1, z_2, \ldots, z_q]^T$, $y \in \mathbb{R}^q$, and $y = [y_1, y_2, \ldots, y_q]^T$, the function $y = \text{sig}^r(z)$ is defined as follows,

$$y_i = \begin{cases} |z_i|^r & \text{if } z_i > 0 \\ 0 & \text{if } z_i = 0 \\ -|z_i|^r & \text{if } z_i < 0 \end{cases} \qquad (3.10)$$

where $|\cdot|$ is the absolute value of real numbers.

The proposed finite-time dual neural network to solve the problem (3.1) is summarized as follows:

$$\text{state equation: } \varepsilon \dot{\mu} = -\text{sig}^r\Big(g\big((I - \rho EME^T)\mu + \rho Es\big) \\ + \rho EME^T \mu - \rho Es\Big) \qquad (3.11\text{a})$$

$$\text{output equation: } x = -ME^T \mu + s \qquad (3.11\text{b})$$

where I is an identity matrix of a proper size, $\varepsilon \in \mathbb{R}$, $\varepsilon > 0$, $\rho \in \mathbb{R}$, $\rho > 0$, $r \in \mathbb{R}$, $0 < r < 1$, M and s are defined in (3.6), the function $g(\cdot)$ and $\text{sig}^r(\cdot)$ are defined in (3.4) and (3.10), respectively.

Remark 3.1 The function family $y = \text{sig}^r(x)$ at different r with $0 < r < 1$ is between the function $y = x$ and $y = \text{sign}(x)$ in values ($y = \text{sign}(x)$ is the sign function, with $y = 1$ for $x > 0$, $y = -1$ for $x < 0$ and $y = 0$ for $x = 0$), as depicted in Fig. 3.1.

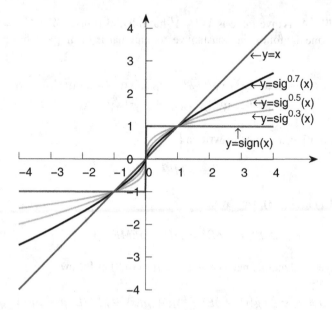

Fig. 3.1 Comparisons of $y = x$, $y = \text{sign}(x)$ and $y = \text{sig}^r x$ at $r = 0.3$, $r = 0.5$, $r = 0.7$

When $r = 1$, the neural network (3.11) reduces to the improved neural network studied in [23]. However, a big difference between our model and the model studied in [23] is that our neural network, for all $0 < r < 1$ (no matter how close to 1 r is), always converges in finite time while the improved neural network never converge in finite time.

Remark 3.2 $y = \text{sig}^r(x)$ is a continuous and smooth function for $0 < r < 1$. It approaches the discontinuous sign function $y = \text{sign}(x)$ when r approaches 0 and it reduces to the sign function when $r = 0$. Due to the discontinuous nature of the sign function, chatter phenomena may happen for the dynamic system with a sign function module in the presence of time delay. To avoid this, the designer may replace the sign function with a piecewise linear saturation function, such as $y = \text{sat}(x)$ with $y = x$ when $-1 < x < 1$, $y = 1$ when $x \geq 1$ and $y = -1$ when $x \leq -1$. However, this replacement sacrifices the fast convergence and accuracy. In [27, 28], a hard-limiting function is used in the recurrent neural network design in order to obtain finite time convergence. The hard-limiting function is in nature similar to the sign function, and therefore is sensitive to possible time delay, and the problem is crucial especially when the neural network is implemented in hardware, in which case time delay is unavoidable. In contrast, by choosing a relative large r with $0 < r < 1$, the proposed neural network has a relative high robustness against time delay and can reach convergence in finite time as well. This point can also be observed in the simulation example in Sect. 3.4.5.

Remark 3.3 Several existing neural network models are available to solve the constrained quadratic programming problem (3.1) [19, 20, 22, 23, 26]. We evaluate them from three different aspects: theoretical error, i.e., whether there exists a steady state error from the output of the neural network to the theoretical solution of the problem, convergence time, i.e., whether it takes finite time or infinite time for convergence, and spatial complexity in terms of the number of neurons included in the neural network. The comparisons of the gradient based neural network [19], lagrange neural network [20], dual neural network [23], simplified dual neural network [26], improved dual neural network [22] (it is noteworthy that the improved dual neural network is only applicable to the case with W being an identity matrix in (3.1)) and the proposed neural network in this chapter are summarized in Table 3.1.

Table 3.1 Comparisons of the proposed model with some existing neural network models for solving Problem (3.1)

Neural network models	Theoretical error	Convergence time	Spatial complexity
Gradient based neural network [19]	Non-zero	Infinite	n
Lagrange neural network [20]	Zero	Infinite	n+m+4q
Dual neural network [23]	Zero	Infinite	m+q
Simplified dual neural network [26]	Zero	Infinite	q
Improved dual neural network [22]	Zero	Infinite	m+q
The proposed neural network	Zero	Finite	q

3.3   Convergence Analysis

In this section, we study the stability of the proposed dynamic neural network, its finite-time convergence property, and the optimality of the result obtained by using this neural network to the original quadratic programming problem. To prove the main results, some equations and inequalities are used. The first one is about the gradient of a general vector norm:

$$\nabla\left(\|x\|_a^a\right) = a \operatorname{sig}^{a-1}(x) \tag{3.12}$$

where ∇ is the vector differential operator, $a \in \mathbb{R}$ and $a > 1$, $x \in \mathbb{R}^n$, $\|x\|_a$ is the a-norm of x, which is defined as $\|x\|_a = \left(\sum_{i=1}^n |x_i|^a\right)^{\frac{1}{a}}$ for $x = [x_1, x_2, \ldots, x_n]^T$. Equation (3.12) can be simply verified by expanding both side of (3.12) in entrywise.

The second useful inequality is as follows:

$$\|x\|_a \leq \|x\|_b \tag{3.13}$$

for $x \in \mathbb{R}^n$, $a \in \mathbb{R}$, $b \in \mathbb{R}$ and $0 < b < a$. This inequality reveals that $\|x\|_a$ is a non-increasing function with respect to a for $a > 0$ and can be simply proven by showing the derivative of $\|x\|_a$ relative to a is always non-positive. Inequality (3.13) establishes the relation between two different norms for a same vector and it can be used to estimate the bound of certain norm of a vector according to knowledge on its other norms.

The following lemma, which is well known as Cauchy interlace theorem, is also useful in the proof.

Lemma 3.1 *Let A be a Hermitian matrix of order n, and let B be a principal sub-matrix of A of order $n - 1$. If $\lambda_n \leq \lambda_{n-1} \leq \cdots \leq \lambda_2 \leq \lambda_1$, lists the eigenvalues of A and $\mu_n \leq \mu_{n-1} \leq \cdots \leq \mu_3 \leq \mu_2$ the eigenvalues of B, then $\lambda_n \leq \mu_n \leq \lambda_{n-1} \leq \mu_{n-1} \leq \cdots \leq \lambda_2 \leq \mu_2 \leq \lambda_1$.*

As a special kind of Hermitian matrices, symmetric matrices also have the property stated in Lemma 3.1. By recursively using Lemma 3.1, we can conclude that the eigenvalues of a symmetric matrix's principal sub-matrix (one or more order less than the symmetric matrix) are greater than or equal to the symmetric matrix's smallest eigenvalue and less than or equal to its greatest eigenvalue.

Another useful lemma is stated as follows:

Lemma 3.2 *([23]) Let $W \in \mathbb{R}^{n \times n}$, $W \succ 0$, $A \in \mathbb{R}^{m \times n}$ $(m < n)$, rank$(A) = m$, $E \in \mathbb{R}^{q \times n}$, then*

$$E(W^{-1} - W^{-1}A^T(AW^{-1}A^T)^{-1}AW^{-1})E^T \succeq 0 \tag{3.14}$$

To our problem, Lemma 3.2 means $EME^T \succeq 0$ with E, and M defined in (3.1) and (3.6), respectively.

The following lemma plays an important role in the proof of the main theorem.

Lemma 3.3 *Let ε_1, ε_q be the smallest and the largest eigenvalues of EME^T, with E, and M defined in (3.1) and (3.6) as $E \in \mathbb{R}^{q \times n}$, $M \in \mathbb{R}^{n \times n}$, $M = M^T$ and let $A_1 = D(I - \rho EME^T) + \rho EME^T$, where I is an identity matrix of proper dimensions, $D \in \mathbb{R}^{q \times q}$, $D = \operatorname{diag}(d_1, d_2, \ldots, d_q)$, with $d_i \in \mathbb{R}$, $0 \le d_i \le 1$ for $i = 1, 2, \ldots, q$, $\rho \in \mathbb{R}$, $0 < \rho \le \frac{2}{\varepsilon_q}$. Then, $A_1 + A_1^T \succeq \rho \varepsilon_1 I$ and $x^T (A_1 + A_1^T) x \ge \rho \varepsilon_1 x^T x$ for $\forall x \in \mathbb{R}^n$. In addition, $ME^T x = 0$ when $x^T (A_1 + A_1^T) x = 0$.*

Proof To show $A_1 + A_1^T \succeq \rho \varepsilon_1 I$, we can equivalently prove $x^T A_1 x \ge \frac{\rho \varepsilon_1 x^T x}{2}$ for $x \in \mathbb{R}^q$. Since $x^T A_1 x$ is affine to d_i for all $i = 1, 2, \ldots, q$, and $0 \le d_i \le 1$, the extremum of $x^T A_1 x - \frac{\rho \varepsilon_1 x^T x}{2}$ is reached when $d_i = 0$ or $d_i = 1$ for $i = 1, 2, \ldots, q$. Accordingly, we only need to prove $x^T A_1 x \ge \frac{\rho \varepsilon_1 x^T x}{2}$ under the condition that $d_i = 0$ or $d_i = 1$ for $i = 1, 2, \ldots, q$. In this situation, D is a diagonal matrix with the diagonal elements to be 0 or 1. To facilitate the analysis, we introduce a permutation transformation matrix, $P \in \mathbb{R}^{q \times q}$ with $PP^T = P^T P = I$, which re-orders the non-zero elements of D to the left upper part of the diagonal. That is,

$$PDP^T = \begin{bmatrix} I_{n_0 \times n_0} & 0 \\ 0 & 0 \end{bmatrix} \qquad (3.15)$$

where n_0 is the number of 1s on the diagonal of D. For the trivial case $n_0 = 0$, which means $D = 0$, the conclusion can be drawn directly by noting that $A_1 = \rho EME^T$ in this case. For the trivial case $n_0 = q$, which means $D = I$, the conclusion is also straightforward since $A_1 = I$ in this case. In the following part, we study the case when $0 < n_0 < q$. In this case,

$$
\begin{aligned}
& x^T A_1 x \\
&= x^T D(I - \rho EME^T) x + \rho x^T EME^T x \\
&= x^T P^T PDP^T P(I - \rho EME^T) P^T Px + \rho x^T P^T PEME^T P^T Px \\
&= (Px)^T (PDP^T) P(I - \rho EME^T) P^T (Px) + \rho (Px)^T (PEME^T P^T)(Px)
\end{aligned}
$$
$$(3.16)$$

Define a new variable $y = Px$ and denote $y_1 \in \mathbb{R}^{n_0}$ the first n_0 elements of y and $y_2 \in \mathbb{R}^{q - n_0}$ the rest $q - n_0$ elements of y, i.e., $y = [y_1^T, y_2^T]^T$. Substituting (3.15) and y_1, y_2, y into (3.16), we get,

$$
\begin{aligned}
& x^T A_1 x \\
&= y^T \begin{bmatrix} I_{n_0 \times n_0} & 0 \\ 0 & 0 \end{bmatrix} P(I - \rho EME^T) P^T y + \rho y^T (PEME^T P^T) y \\
&= [y_1^T \ 0] (I - \rho PEME^T P^T) \begin{bmatrix} y_1 \\ y_2 \end{bmatrix} + \rho y^T (PEME^T P^T) y \\
&= y_1^T y_1 - \rho [y_1^T \ 0] (PEME^T P^T) \begin{bmatrix} y_1 \\ y_2 \end{bmatrix} + \rho y^T (PEME^T P^T) y \qquad (3.17)
\end{aligned}
$$

Define a symmetric matrix $A_2 \in \mathbb{R}^{q \times q}$ and $A_2 = PEME^T P^T$. A_2 has the same eigenvalues as EME^T since A_2 and EME^T are similar matrices by noting $P^T = P^{-1}$. Partitioning A_2 into blocks $A_2 = \begin{bmatrix} B_1 & B_2 \\ B_2^T & B_3 \end{bmatrix}$ with $B_1 \in \mathbb{R}^{n_0 \times n_0}$, $B_1 = B_1^T$, $B_3 \in \mathbb{R}^{(q-n_0) \times (q-n_0)}$, $B_3 = B_3^T$, $B_2 \in \mathbb{R}^{n_0 \times (q-n_0)}$, we get,

$$
\begin{aligned}
x^T A_1 x \\
&= y_1^T y_1 - \rho \begin{bmatrix} y_1^T & 0 \end{bmatrix} \begin{bmatrix} B_1 & B_2 \\ B_2^T & B_3 \end{bmatrix} \begin{bmatrix} y_1 \\ y_2 \end{bmatrix} + \rho \begin{bmatrix} y_1^T & y_2^T \end{bmatrix} \begin{bmatrix} B_1 & B_2 \\ B_2^T & B_3 \end{bmatrix} \begin{bmatrix} y_1 \\ y_2 \end{bmatrix} \\
&= y_1^T y_1 + \rho y_1^T B_2 y_2 + \rho y_2^T B_3 y_2 \\
&= y_1^T y_1 + \rho y_1^T B_2 y_2 + \rho y_2^T B_3 y_2 - \frac{\rho y^T A_2 y}{2} + \frac{\rho y^T A_2 y}{2} \\
&= y_1^T y_1 + \rho y_1^T B_2 y_2 + \rho y_2^T B_3 y_2 - \frac{\rho}{2} (y_1^T B_1 y_1 + y_2^T B_3 y_2 + 2 y_1^T B_2 y_2) + \frac{\rho}{2} y^T A_2 y \\
&= y_1^T (I - \frac{\rho B_1}{2}) y_1 + \frac{\rho y_2^T B_3 y_2}{2} + \frac{\rho y^T A_2 y}{2}
\end{aligned}
\tag{3.18}
$$

Since ε_1, ε_q are the smallest and the largest eigenvalues of EME^T, respectively, ε_1, ε_q are also the smallest and the largest eigenvalues of A_2, respectively (As stated above, EME^T has the same eigenvalues as A_2). Therefore, $\varepsilon_1 I \preceq A_2 \preceq \varepsilon_q I$. According to Lemma 3.1, B_1, B_3, which are principal sub-matrices of A_2, have all eigenvalues less than or equal to ε_q and greater than or equal to ε_1, i.e., $\varepsilon_1 I \preceq B_1 \preceq \varepsilon_q I$ and $\varepsilon_1 I \preceq B_3 \preceq \varepsilon_q I$. Based on this analysis, we get,

$$
\begin{aligned}
x^T A_1 x \\
&\geq (1 - \frac{\rho \varepsilon_q}{2}) y_1^T y_1 + \frac{\rho \varepsilon_1 y_2^T y_2}{2} + \frac{\rho \varepsilon_1 y^T y}{2} \\
&\geq \frac{\rho \varepsilon_1 y^T y}{2} \\
&= \frac{\rho \varepsilon_1 (P^T x)^T (P^T x)}{2} \\
&= \frac{\rho \varepsilon_1 x^T x}{2}
\end{aligned}
\tag{3.19}
$$

Note that $P^T = P^{-1}$ is used in the derivation of (3.19).

As to the fact that $E^T x = 0$ when the equality in $x^T (A_1 + A_1^T) x = 0$ holds, it can be proved by noticing the following result for $x^T (A_1 + A_1^T) x = 0$ from (3.18),

$$
y_1^T (I - \frac{\rho B_1}{2}) y_1 + \frac{\rho y_2^T B_3 y_2}{2} + \frac{\rho y^T A_2 y}{2} = 0
\tag{3.20}
$$

Since $I - \frac{\rho B_1}{2} \succeq 0$, $B_3 + B_3^T \succeq 0$ and $A_2 \succeq 0$, we obtain the following,

$$y_1^T (I - \frac{\rho B_1}{2}) y_1 = 0 \tag{3.21a}$$

$$\frac{\rho y_2^T B_3 y_2}{2} = 0 \tag{3.21b}$$

$$\frac{\rho y^T A_2 y}{2} = 0 \tag{3.21c}$$

Clearly, (3.21c) implies $A_2 y = 0$ since A_2 is symmetric and semi-positive definite. Recalling that $A_2 y = PEME^T P^T P x = PEME^T x$ and P has a full rank, we thus conclude that $A_2 y = 0$ results in $EME^T x = 0$. Also, $EME^T x = 0$ is equivalent to $x^T EME^T x = 0$ since $EME^T \succeq 0$. Noticing that $x^T EME^T x = (E^T x)^T M (E^T x)$ and $M \succeq 0$, we conclude that $ME^T x = 0$. This completes the proof.

Now we are on the stage to present the main theorem:

Theorem 3.1 *Let ε_1 and ε_q be the smallest and the largest eigenvalues of EME^T, respectively (according to Lemma 3.2, $\varepsilon_q > \varepsilon_1 \geq 0$), the neural network (3.11) with $\varepsilon > 0$, $0 < r < 1$ and $0 < \rho \leq \frac{2}{\varepsilon_q}$ is stable in the sense of Lyapunov. Moreover, the neural network converges if μ_0, which is the solution of $g\left((I - \rho EME^T)\mu_0 + \rho Es\right) + \rho EME^T \mu_0 - \rho Es = 0$, is inside the largest invariant set constructed by the system dynamics (3.11) and the constraint $ME^T \operatorname{sig}^r \left(g\left((I - \rho EME^T)\mu' + \rho Es\right) + \rho EME^T \mu' - \rho Es\right) = 0$ in terms of μ'. In addition, if EME^T has full rank, the neural network converges to an equilibrium point μ^* in finite time and the convergence time is not longer than $\frac{2\varepsilon \left\| g\left((I - \rho EME^T)\mu_0 + \rho Es\right) + \rho EME^T \mu_0 - \rho Es \right\|_{r+1}^{1-r}}{\rho \varepsilon_1 (1 - r)}$.*

Proof To prove the conclusion, we construct the following Lyapunov function,

$$V = \frac{\left\| g\left((I - \rho EME^T)\mu + \rho Es\right) + \rho EME^T \mu - \rho Es \right\|_{r+1}^{r+1}}{r+1} \tag{3.22}$$

where $\|x\|_{r+1}$ denotes the $r+1$ norm of a vector x. That is, we have $\|x\|_{r+1} = \left(\sum_{i=1}^n |x_i|^{r+1}\right)^{\frac{1}{r+1}}$ for $x = [x_1, x_2, \ldots, x_n]^T$. Recalling Eq. (3.12), the time derivative of V along the neural network trajectory (3.11a) can be obtained as follows:

$$\dot{V} = \dot{\mu}^T \left(J(I - \rho EME^T) + \rho EME^T\right)^T$$
$$\cdot \operatorname{sig}^r \left(g\left((I - \rho EME^T)\mu + \rho Es\right) + \rho EME^T \mu - \rho Es\right)$$
$$= -\frac{1}{\varepsilon}\left(\operatorname{sig}^r \left(g\left((I - \rho EME^T)\mu + \rho Es\right)\right. \right.$$
$$\left. \left. + \rho EME^T \mu - \rho Es\right)\right)^T \left(J(I - \rho EME^T)\right)$$

$$+ \rho EME^T)^T \text{sig}^r \Big(g\big((I - \rho EME^T)\mu$$
$$+ \rho Es\big) + \rho EME^T \mu - \rho Es\Big) \tag{3.23}$$

where $J = D^+ g$ is the upper-right dini-derivative of $g\big((I - \rho EME^T)\mu + \rho Es\big)$. According to the definition of $g(\cdot)$ in (3.4), we know J is a diagonal matrix $J = \text{diag}(J_1, J_2, \ldots, J_n)$ and the ith diagonal element J_i is as follows,

$$J_i = \begin{cases} 1 & \text{if } \rho l_i \leq \big((I - \rho EME^T)\mu + \rho Es\big)_i < \rho h_i \\ 0 & \text{if } \big((I - \rho EME^T)\mu + \rho Es\big)_i < \rho l_i \text{ or} \\ & \big((I - \rho EME^T)\mu + \rho Es\big)_i \geq \rho h_i \end{cases} \tag{3.24}$$

where l_i, h_i are as defined in (3.4), $\big((I - \rho EME^T)\mu + \rho Es\big)_i$ is the ith element of the vector $(I - \rho EME^T)\mu + \rho Es$. According to Lemma 3.3, $\big(J(I - \rho EME^T) + \rho EME^T\big)^T + \big(J(I - \rho EME^T) + \rho EME^T\big) \succeq \rho \varepsilon_1 I$ with ε_1 denoting the smallest eigenvalue of EME^T. Bringing this inequality into (3.23) yields,

$$\dot{V} \leq -\frac{\rho \varepsilon_1}{2\varepsilon} \left\| \text{sig}^r \Big(g\big((I - \rho EME^T)\mu + \rho Es\big) + \rho EME^T \mu - \rho Es\Big) \right\|^2$$
$$= -\frac{\rho \varepsilon_1}{2\varepsilon} \left\| g\big((I - \rho EME^T)\mu + \rho Es\big) + \rho EME^T \mu - \rho Es \right\|_{2r}^{2r} \tag{3.25}$$

where $\|\cdot\|$, $\|\cdot\|_{2r}$ represent 2-norm and 2r-norm, respectively. The equality in (3.25) can be verified by expanding the vector in $\|\cdot\|$ into elements in every dimensions. Noting that $r + 1 > 2r > 0$ for $0 < r < 1$, according to the norm inequality (3.13), we further get,

$$\dot{V} \leq -\frac{\rho \varepsilon_1}{2\varepsilon} \left\| g\big((I - \rho EME^T)\mu + \rho Es\big) + \rho EME^T \mu - \rho Es \right\|_{r+1}^{2r}$$
$$= -\frac{\rho \varepsilon_1}{2\varepsilon} \left(\big((r+1)V\big)^{\frac{1}{r+1}} \right)^{2r}$$
$$= -\frac{\rho \varepsilon_1}{2\varepsilon} (r+1)^{\frac{2r}{r+1}} V^{\frac{2r}{r+1}} \tag{3.26}$$

According to Lemma 3.2, $EME^T \succeq 0$ and therefore $\varepsilon_1 \geq 0$. Accordingly, $\dot{V} \leq 0$, which proves the stability of the neural network (3.11).

Particularly if EME^T has full rank, there will be no zero eigenvalue for EME^T. In this case, $EME^T \succ 0$ and $\varepsilon_1 > 0$. To prove the finite-time convergence under $\varepsilon_1 > 0$, we first construct an auxiliary scalar dynamic system $\dot{K} = -\frac{\rho \varepsilon_1}{2\varepsilon}(r+1)^{\frac{2r}{r+1}} K^{\frac{2r}{r+1}}$ with the following initial value of K,

$$K_0 = V_0 = V(t_0) = \frac{\left\| g\big((I - \rho EME^T)\mu_0 + \rho Es\big) + \rho EME^T \mu_0 - \rho Es \right\|_{r+1}^{r+1}}{r+1}$$

with $\mu_0 = \mu(t_0)$. The solution of the auxiliary system can be solved by separation of variables. The solution is,

$$K(t) = \begin{cases} \left(V_0^{\frac{1-r}{r+1}} - \Delta(t - t_0)\right)^{\frac{r+1}{1-r}} & \text{when } t_0 \le t < t_1 \\ 0 & \text{when } t \ge t_1 \end{cases} \quad (3.27)$$

where $\Delta \in \mathbb{R}$, $\Delta = \frac{\rho\varepsilon_1}{2\varepsilon}(1-r)(r+1)^{\frac{r-1}{r+1}}$, is a constant. $t_1 \in \mathbb{R}, t_1 \ge 0, t_1 = t_0 + \frac{V_0^{\frac{1-r}{r+1}}}{\Delta}$. According to the Comparison Lemma, we know $V(t) \le K(t)$ for all $t \ge t_0$. Therefore, we get,

$$\begin{cases} V(t) \le \left(V_0^{\frac{1-r}{r+1}} - \Delta(t - t_0)\right)^{\frac{r+1}{1-r}} & \text{when } t_0 \le t < t_1 \\ V(t) = 0 & \text{when } t \ge t_1 \end{cases} \quad (3.28)$$

With the definition of V in (3.22), we know that $V(t) = 0$ when $t \ge t_1$ suggests $g\left((I - \rho EME^T)\mu + \rho Es\right) + \rho EME^T\mu - \rho Es = 0$, which means $\mu = \mu^*$ since μ^* is an equilibrium point satisfying $g\left((I - \rho EME^T)\mu^* + \rho Es\right) + \rho EME^T\mu^* - \rho Es = 0$. Accordingly, we conclude that the proposed neural network converges to an equilibrium point μ^* in finite time, which is not greater than $t_1 - t_0 = \frac{2\varepsilon V_0^{\frac{1-r}{r+1}}}{\rho\varepsilon_1(1-r)(r+1)^{\frac{r-1}{r+1}}} = \frac{2\varepsilon \left\| g\left((I-\rho EME^T)\mu_0 + \rho Es\right) + \rho EME^T\mu_0 - \rho Es \right\|^{1-r}_{r+1}}{\rho\varepsilon_1(1-r)}$.

This completes the proof.

The following Theorem reveals the relation between the equilibrium point of the neural network and the optimal solution to the quadratic programming problem (3.1).

Theorem 3.2 *Let μ^* be an equilibrium point of (3.11a), the output of the finite-time dual neural network at μ^*, which is $x^* = -\rho ME^T\mu^* + s$, is the optimal solution to the quadratic programming problem (3.1).*

Proof Note that the equilibrium point μ^* satisfies $\text{sig}^r\left(g\left((I - \rho EME^T)\mu^* + \rho Es\right) + \rho EME^T\mu^* - \rho Es\right) = 0$, i.e., $g\left((I - \rho EME^T)\mu^* + \rho Es\right) + \rho EME^T\mu^* - \rho Es = 0$, which means μ^* is also a solution to Eq. (3.8). Thus, (λ^*, μ^*, x^*) is a solution to the equation set composed of Eqs. (3.7), (3.5b), and (3.8), where $\lambda^* = -\rho(AW^{-1}A^T)^{-1}AW^{-1}E^T\mu^* - (AW^{-1}A^T)^{-1}(b + AW^{-1}c)$. Due to the equivalence of Eq. (3.2) and the equation set composed of (3.7), (3.5b), and (3.8), we know (λ^*, μ^*, x^*) is also a solution to Eq. (3.2). Since the solution to Eq. (3.2), according to KKT conditions, is the optimal solution to the quadratic programming problem (3.1) in dual space, we conclude that x^* is the optimal solution of (3.1), which completes the proof.

3.4   An Illustrative Example

In this section, we solve a simulation example to illustrate the performance of the proposed neural network (3.11). We consider the following quadratic programming problem:

$$\text{minimize } 3x_1^2 + 3x_2^2 + 4x_3^2 + 5x_4^2 + 3x_1x_2 + 5x_1x_3 + x_2x_4 - 11x_1 - 5x_4$$
$$\text{subject to } 3x_1 - 3x_2 - 2x_3 + x_4 = 0$$
$$4x_1 + x_2 - x_3 - 2x_4 = 0$$
$$-73 \le -50x_1 + 50x_2 \le -50$$
$$-20 \le 32x_1 + 10x_3 \le 41$$

$$(3.29)$$

In this problem, parameters are as follows:

$$W = \begin{bmatrix} 6 & 3 & 5 & 0 \\ 3 & 6 & 0 & 1 \\ 5 & 0 & 8 & 0 \\ 0 & 1 & 0 & 10 \end{bmatrix}, \quad c = \begin{bmatrix} -11 \\ 0 \\ 0 \\ -5 \end{bmatrix}, \quad A = \begin{bmatrix} 3 & -3 & -2 & 1 \\ 4 & 1 & -1 & -2 \end{bmatrix}, \quad b = \begin{bmatrix} 0 \\ 0 \end{bmatrix}$$

$$E = \begin{bmatrix} -50 & 50 & 0 & 0 \\ 32 & 0 & 10 & 0 \end{bmatrix}, \quad l = \begin{bmatrix} -73 \\ -20 \end{bmatrix}, \quad h = \begin{bmatrix} -50 \\ 41 \end{bmatrix} \quad (3.30)$$

For this problem, the largest eigenvalue of the matrix EME^T is $\varepsilon_q = 138.43$. We choose $\rho = 0.01$, which satisfies $0 < \rho \le \frac{2}{\varepsilon_q}$. In addition, the matrix EME^T has full rank, so the neural network converges to the optimal solution in finite time according to Theorem 3.1. The scaling factor ε is chosen to be 10^{-8} in the simulation. In this section, we use this illustrative example to systematically evaluate the performance of the proposed neural network in four aspects: accuracy, convergence speed, sensitivity to additive noise and robustness against time delay.

3.4.1   Accuracy

We use $r = 0.6$ as a particular example to demonstrate the accuracy of our method (simulations with other values of r and the comparison between different r are performed in the following sections). Simulation is run for 7×10^{-8} s and the simulation results in Figs. 3.2 and 3.3 show the evolution of μ and x in this period. At the end of the simulation, the neural network output is $x = [0.4999999992, -0.4999999986, 1.4999999970, 0.0000000007]$, which is very close to the theoretical optimal solution $x = [0.5, -0.5, 1.5, 0]$. Note that there is a tiny error between the output of the neural network and the theoretical optimal solution. This error results from truncation, approximation, limited step size, etc. in numerical simulations. Figures 3.4 and 3.5 show the trajectory of x_1 and x_2 and that of x_3 and x_4 from different initial

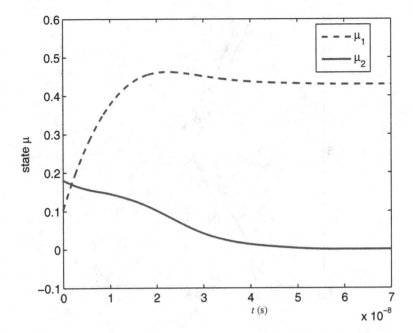

Fig. 3.2 Transient behavior of μ in the illustrative example in Sect. 3.4

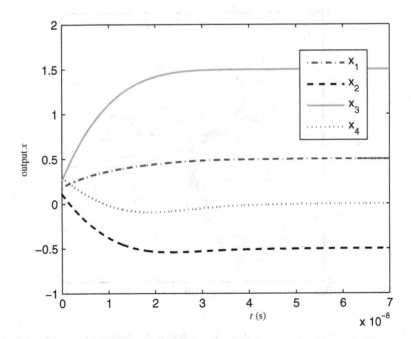

Fig. 3.3 Transient behavior of x in the illustrative example in Sect. 3.4

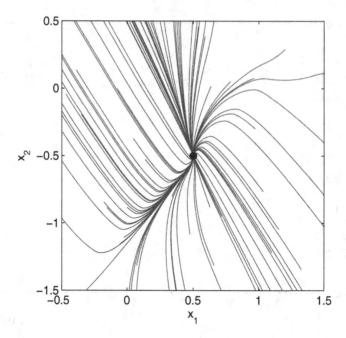

Fig. 3.4 Trajectory of x_1 and x_2 from different initial states in the illustrative example in Sect. 3.4

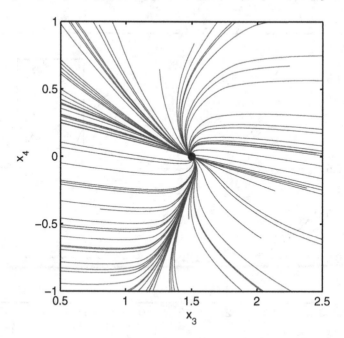

Fig. 3.5 Trajectory of x_3 and x_4 from different initial states in the illustrative example in Sect. 3.4

states, which verifies the convergence property of the neural network under different initializations.

3.4.2 Convergence Speed

In this section, we compare the convergence speed of the proposed neural network at different values of r. As shown in Theorem 3.1, the convergence time has dependence on the initialization μ_0. To eliminate the effect of random initialization, results are averaged over 100 Monte Carlo runs with independently generated initialization of Gaussian distribution with zero means and unit variance. To evaluate the performance, the evolution of errors measured in 2-norm under a set of r are plotted in Fig. 3.6. From this figure, we can see that the neural network with a larger r has a faster convergence at the beginning of the simulation (see the small window). With the elapse of time, the neural network with a smaller r gradually surpasses others in convergence. At the end of the simulation, the errors reach zero for $r = 0, r = 0.2, r = 0.4, r = 0.6, r = 0.8$, which verifies the finite-time convergence property of the proposed neural network model. In contrast, the neural network with $r = 1$ still has a gap from the zero error at the end.

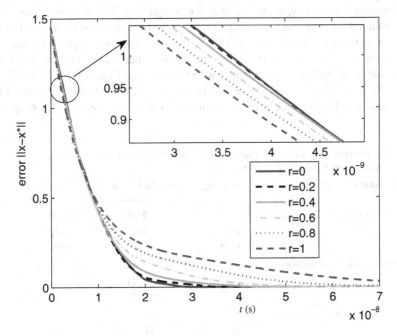

Fig. 3.6 Comparisons of errors under $r = 0, r = 0.2, r = 0.4, r = 0.6, r = 0.8, r = 1$ in the illustrative example in Sect. 3.4

3.4.3   Comparisons on Computational Efficiency in Numerical Simulations

The proposed neural network is described in continuous-time ordinary differential equations and is implementable on parallel computing devices, such as analog circuits, with the ideal convergence speed. However, when it comes to the simulation on a digital computer, the neural network dynamics must be firstly discretized into an iteration scheme. We use the first-order Euler method to discretize the dynamic neural network model (3.11) with a fixed time-step length. Then, the discrete-time version of the proposed model is simulated and compared with the neural network models proposed in recent years to solve the constrained quadratic programming problem, namely, the dual neural network [23], the simplified dual neural network (it is noteworthy that the improved dual neural network [22] only solve the quadratic programming problem with W being an identity matrix and does not apply to the problem (3.29)). For both the dual neural network and the simplified dual neural network, Euler method is employed to discretize the neuro-dynamics. For the three compared neural network models, the scaling factor is chosen as $\varepsilon = 10^{-8}$. For the proposed neural network model, we simply choose $r = 0.6$ and $\rho = 1$ for the simulation. For the simplified dual neural network and the proposed neural network, the time-step length is chosen as 10^{-10} s. The dual neural network diverges with this step length and a smaller time step 10^{-12} s, which compromises the convergence and computational efficiency, is employed in the comparison. The stopping criteria for the neural networks is that the error $\|x - x^*\|$ (x^* is the theoretical optimal solution) is lower than a predefined threshold (10^{-2}, 10^{-3} and 10^{-4} are respectively used as the threshold in the comparisons). The simulation is performed with the programming language Matlab 7.8 on a laptop with the Intel (R) Core(TM) 2 Duo CPU at 1.80 GHz and 2GB of RAM. Table 3.2 shows the comparison results averaged by running Monte Carlo simulations for 50 times. As shown in the table, the proposed

Table 3.2 Comparisons on Computational Efficiency in Numerical Simulations of the proposed model with some existing neural network models

Neural network models	Stopping criteria	CPU time (seconds)	Number of iterations
Dual neural network [22]	10^{-2}	0.2571	3205
	10^{-3}	0.4235	4512
	10^{-4}	0.6304	6226
Simplified dual neural network [26]	10^{-2}	0.0552	403
	10^{-3}	0.0611	627
	10^{-4}	0.0856	911
The proposed neural network	10^{-2}	0.0442	121
	10^{-3}	0.0507	188
	10^{-4}	0.0536	221

neural network outperforms its counterparts in the sense of both the CPU time and the number of iterations.

3.4.4 Sensitivity to Additive Noise

In practice, noise may pollute the dynamics of the neural network. Especially when the neural network is implemented in analog circuits, additive noise is often unavoidable. In this part, we compare the sensitivity of the neural network to additive noise under different r. For simplicity, we only consider the presence of noise in the state equation. That is, we consider the following neural dynamics:

$$\varepsilon\dot{\mu} = -\text{sig}^r\left(g\big((I - \rho EME^T)\mu + \rho Es\big) + \rho EME^T\mu - \rho Es\right) + v \quad (3.31a)$$

$$x = -ME^T\mu + s \quad (3.31b)$$

where v is zero mean Gaussian white noise with covariance σI. As did before, results are averaged over 100 Monte Carlo runs with independently generated initialization of Gaussian distribution with zero means and unit variance to eliminate the effect of random initialization. Figures 3.7 and 3.8 show the 2-norm of the difference between the output of the neural network x and the theoretical optimal point x^* under different r and at different noise level of σ. From these figures, we can see that the output of the neural network cannot reach the ideal optimal solution of the original optimization problem and the output demonstrates a certain of randomness both due to the presence of noise. However, the neural network is still able to output an approximation of the optimal solution and the accuracy of this approximation depends on the noise level σ. It can be observed in Figs. 3.7 and 3.8 that the error increases, in statistical sense, with the increase of σ, which is the noise level and on the other hand, the neural network with a smaller r is less sensitive to the additive noise under a given σ.

3.4.5 Robustness Against Time Delay

In the ideal model of the proposed neural network, time delay is not taken into account. However, in implementation of the neural network, such as the implementation with analog circuits, time delay is inevitable due to limited response rate and sometimes it is crucial to the stability of the system. With this consideration, in this part, we evaluate the influence of time delay on the neural computing with the proposed neural network under different values of r. We consider the time delay in the feedback channel of the state equation as follows:

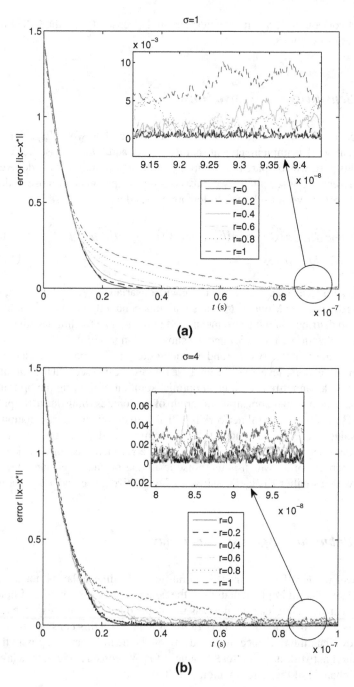

Fig. 3.7 Comparisons of errors under $r = 0, r = 0.2, r = 0.4, r = 0.6, r = 0.8, r = 1$ with noise level $\sigma = 1, \sigma = 4$ in the illustrative example in Sect. 3.4

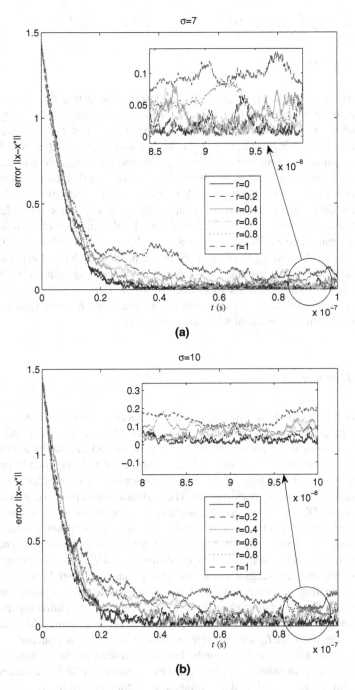

Fig. 3.8 Comparisons of errors under $r = 0$, $r = 0.2$, $r = 0.4$, $r = 0.6$, $r = 0.8$, $r = 1$ with noise level $\sigma = 7$, $\sigma = 10$ in the illustrative example in Sect. 3.4

$$\varepsilon\dot{\mu}(t) = -\operatorname{sig}^r\left(g\left((I - \rho EME^T)\mu(t - \phi) + \rho Es\right) + \rho EME^T\mu(t - \phi) - \rho Es\right)$$
$$\tag{3.32a}$$

$$x(t) = -ME^T\mu(t) + s \tag{3.32b}$$

where ϕ is the time delay, $\mu(t)$, $\mu(t - \phi)$, and $x(t)$ are the state at time t, the state at time $t - \phi$ and the output at time t of the neural network, respectively. For $-\tau \le t \le 0$, x is initialized to be a same value, i.e., $x(t) = x(0)$ for $-\tau \le t \le 0$. Since ε is a convergence rate scaling factor, we use $\frac{1}{\varepsilon}$ as the time unit. The simulation is performed with a zero mean and unit variance initialization, and the results are averaged over 100 Monte Carlo runs. From Figs. 3.9 and 3.10, we can see that the neural network converges smoothly when the time delay is small (see Fig. 3.9a). With the increase of the time delay, the neural network with a relative small r starts to oscillate (see Figs. 3.9b and 3.10a for $r = 0$, $r = 0.2$, $r = 0.4$.). Further increase of the time delay makes the neural network to oscillate under $r = 0$, $r = 0.2$, $r = 0.4$, $r = 0.6$, $r = 0.8$ (see Fig. 3.10b). Among all of them, the neural network with $r = 1$ is most robust to time delay and it demonstrates an oscillation with decaying amplitude when the time delay is τ, while neural networks under $r = 0$, $r = 0.2$, $r = 0.4$, $r = 0.6$, $r = 0.8$ demonstrate oscillations without attenuation. The neural network with a smaller r is more subjective to oscillate when delay appears.

3.4.6 Discussion

This illustrative example on one hand verifies the accuracy and the finite-time convergence property of the proposed neural network under $0 < r < 1$ and it also provides a statistical comparison on the convergence speed, sensitivity to noise and robustness against time delay with different choices of r. In practice, a proper r should be selected by trading-off the above mentioned factors in order to obtain a compressively satisfactory performance. Theoretical and quantitative exploration on the resistance of the proposed model to additive noise or time delay would be helpful for real applications. On the issue of the robustness of the proposed model against additive noise, it is noteworthy that Theorem 3.1 implies the proposed neural network model is exponentially stable under the condition that EME^T has full rank as in this case μ is finite-time convergent and we can always find an enough large $a_0 > 0$ and an enough small $a_1 > 0$ such that $\|\mu(t) - \mu^*\| < a_0 e^{-a_1 t}$ (this inequality only take effects before the time point when $\mu = \mu^*$ and it holds unconditionally after this critical time). The exponential stability of system (3.11), with some additional mild conditions, results in the L_2 stability of the system (3.31) with v as input and $x - x^*$ (x^* is the ideal output with $v = 0$) as output. This result reveals the fact that the additive noise v in (3.31) with finite energy will generate an output x with bounded deviation from x^*. On the issue of time delay robustness, sufficient conditions expressed as a set of linear matrix inequalities (LMIs) can be obtained to guarantee the stability of a cellular neural network with delay. Inspired by this result, it is expectable to

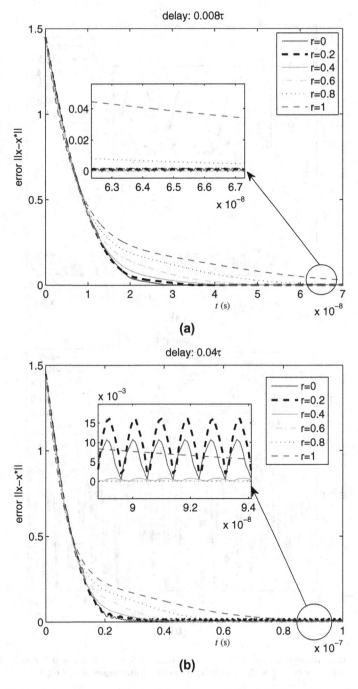

Fig. 3.9 Comparisons of errors under $r = 0$, $r = 0.2$, $r = 0.4$, $r = 0.6$, $r = 0.8$, $r = 1$ with time delay equal to 0.008τ, 0.04τ ($\tau = \frac{1}{\varepsilon}$ is the time unit) in the illustrative example in Sect. 3.4

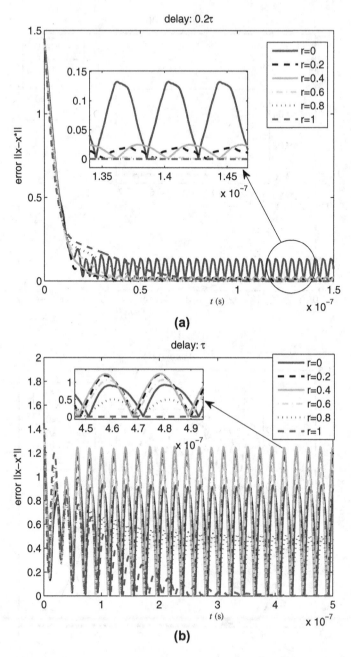

Fig. 3.10 Comparisons of errors under $r = 0$, $r = 0.2$, $r = 0.4$, $r = 0.6$, $r = 0.8$, $r = 1$ with time delay equal to 0.2τ, τ ($\tau = \frac{1}{\varepsilon}$ is the time unit) in the illustrative example in Sect. 3.4

obtain a set of LMIs for system (3.32) to guarantee its the stability and convergence under time delay. However, the strong nonlinearity in the proposed neural network poses great challenges to the quantitive analysis of the above mentioned two aspects, and makes it nontrivial to conduct such an analysis. The two problems are still open questions and will be investigated in our future work.

3.5 Solving k-WTA with the Proposed Neural Network

In this section, we apply the proposed finite-time dual neural network to solve the k-WTA problem and tailor the neural network to that with only a single neuron by exploiting the specialty of the k-WTA problem [22, 31]. In contrast to existing works on k-WTA problem solving by using infinite-time convergence recurrent neural networks, and those using hard-limiting activation function to gain finite-time convergence, our method has finite convergence time, the activation function is continuous and the neural network has a free choosing parameter r ($0 < r < 1$) to trade-off convergence speed, sensitivity to additive noise and robustness to possible time delay. In this section, we first formulate the k-WTA problem as a quadratic programming problem and then give theoretical analysis on the stability, finite-time convergence and optimality of the proposed neural network for solving k-WTA. At the end this section, we give simulations to show the effectiveness of our method.

3.5.1 Quadratic Programming Formulation for k-TWA

The k-WTA problem is defined to be the following mapping:

$$x_i = f(u_i) = \begin{cases} 1 & u_i \in \{k \text{ largest elements of u}\} \\ 0 & \text{otherwise} \end{cases} \tag{3.33}$$

k-WTA problem can be equivalently translated into a linear programming problem or a quadratic programming problem. As to the quadratic programming formulation, it has the following form [23]:

$$\text{minimize} \quad \alpha x^T x - u^T x \tag{3.34a}$$
$$\text{subject to} \quad \sum_{i=1}^{n} x_i = k \tag{3.34b}$$
$$0 \leq x_i \leq 1 \quad \text{for } i \in \mathbb{S} \tag{3.34c}$$

where $\mathbb{S} = \{1, 2, \ldots, n\}$, $\alpha \in \mathbb{R}$ and $\alpha > 0$. It has been proven that the optimal solution to the quadratic programming problem (3.34) is also the solution to the k-WTA problem under the condition that the kth largest element (denoted by $\bar{u}_k$) in

u is strictly larger than the $(k + 1)$th largest one (denoted by $\bar{u}_{k+1}$) and the constant parameter α satisfies $\alpha \leq \frac{\bar{u}_k - \bar{u}_{k+1}}{2}$ [23].

Conventionally, researchers use either discrete time iterative methods or continuous time recurrent neural network methods to solve the equivalent linear programming or the quadratic programming to get the solution of the k-WTA problem. To a large-scale problem, in which the dimension of u is very large or even huge, it takes a long computational time for iterative methods to converge due to the time complexity of the algorithms. Recurrent neural network methods, thank to its parallel nature, has a much faster convergence. However, even this, most early works using recurrent neural networks for k-WTA problem solving, in theory, take infinite time to output the ideal optimal solution. Noticing this drawback, neural networks with a hard-limiting activation function are developed to gain finite-time convergence [22]. In the following part, we will show that the finite-time convergence property can still be reached with a different type of activation function in the neural network for solving k-WTA.

Problem (3.34) is a special case of the quadratic programming problem (3.1) by choosing $W = 2\alpha I, c = -u, A = \mathbf{1}^T, b = k, E = I, l = \mathbf{0}, h = \mathbf{1}$ with I to be a $n \times n$ identity matrix, $\mathbf{0}$ and $\mathbf{1}$ to be a n-dimensional vector with all elements equal to zero and a n-dimensional vector with all elements equal to one. According to Theorems 3.1 and 3.2, we can solve the problem (3.34) by using the following neural network,

$$\varepsilon \dot{\mu} = - \text{sig}^r \left(g\big((I - \rho M)\mu + \rho s\big) + \rho M \mu - \rho s \right) \tag{3.35a}$$

$$x = -M\mu + s \tag{3.35b}$$

where I is an identity matrix of a proper size, $\varepsilon \in \mathbb{R}, \varepsilon > 0, \rho \in \mathbb{R}, \rho > 0,$ $r \in \mathbb{R}, 0 < r < 1, M = \frac{1}{2\alpha}(I - \frac{\mathbf{1}\mathbf{1}^T}{n}), s = \frac{1(2\alpha k - \mathbf{1}^T u)}{2\alpha n} + \frac{u}{2\alpha}$ with $\mathbf{1}$ denoting a vector with all entries identical to 1, the function $g(\cdot)$ and $\text{sig}^r(\cdot)$ are defined in (3.4) and (3.10), respectively. This neural network for solving the k-WTA problem needs in total n neurons. Actually, by exploiting the specialty of the problem (3.34), whose inequalities are all box constraints and every quadratic term in the cost function has an equal weight, problem (3.34) can be solved using a single dynamic neuron in finite time. To see this, we equivalently write (3.34) into the following form,

$$\text{minimize} \quad \frac{\rho}{2}x^T x - \frac{\rho}{2\alpha}u^T x \tag{3.36a}$$

$$\text{subject to} \quad \sum_{i=1}^{n} x_i = k \tag{3.36b}$$

$$0 \leq x_i \leq 1 \quad \text{for } i \in \mathbb{S} \tag{3.36c}$$

where $\rho \in \mathbb{R}, \rho > 0$. According to the KKT condition, we get,

$$\rho x - \frac{\rho u}{2\alpha} + \mathbf{1}\lambda + \mu = 0 \tag{3.37a}$$

$$\mathbf{1}^T x = k \tag{3.37b}$$

$$\begin{cases} x = h & \text{if } \mu > 0 \\ l < x < h & \text{if } \mu = 0 \\ x = l & \text{if } \mu < 0 \end{cases} \tag{3.37c}$$

where $\lambda \in \mathbb{R}$ and $\mu \in \mathbb{R}^n$ are dual variables to the equality constraint (3.36b) and the box constraint (3.36c), respectively. Introducing the saturation function $g(\cdot)$ as defined in (3.4), (3.37c) can be simplified to

$$\rho x = g(\rho x + \mu) \tag{3.38}$$

Note that $\rho x + \mu = \frac{\rho u}{2\alpha} - \mathbf{1}\lambda$ according to (3.37a), so we can get the expression of x in terms of λ from (3.38), which writes,

$$x = \frac{1}{\rho} g\left(\frac{\rho u}{2\alpha} - \mathbf{1}\lambda\right) \tag{3.39}$$

Similarly, with (3.37a) and (3.39), we can express μ in terms of λ as follows:

$$\mu = -\lambda\mathbf{1} + \frac{\rho u}{2\alpha} - g\left(\frac{\rho u}{2\alpha} - \lambda\mathbf{1}\right) \tag{3.40}$$

With (3.37b) and (3.39), we get,

$$k = \frac{1}{\rho}\mathbf{1}^T g\left(\frac{\rho u}{2\alpha} - \lambda\mathbf{1}\right) \tag{3.41}$$

We use the following dynamic neuron to solve (3.41) in finite time:

$$\varepsilon\dot{\lambda} = \text{sig}^r\left(\mathbf{1}^T g(\frac{\rho u}{2\alpha} - \lambda\mathbf{1}) - \rho k\right) \tag{3.42}$$

where $\varepsilon \in \mathbb{R}$, $\varepsilon > 0$ is a scaling factor.

To summarize, we use the following neural network with a single dynamic neuron to solve the k-WTA problem (3.33) (equivalent to the quadratic programming problem (3.34)) in finite time:

state equation: $\quad \varepsilon\dot{\lambda} \qquad = \text{sig}^r\left(\mathbf{1}^T g(\frac{\rho u}{2\alpha} - \lambda\mathbf{1}) - \rho k\right) \tag{3.43a}$

output equation: $\quad x \qquad = \frac{1}{\rho} g\left(\frac{\rho u}{2\alpha} - \mathbf{1}\lambda\right) \tag{3.43b}$

where I is an $n \times n$ identity matrix, $\mathbf{1}$ is a n dimensional vector with all entries equal to 1, $\varepsilon \in \mathbb{R}$, $\varepsilon > 0$, $\rho \in \mathbb{R}$, $\rho > 0$, $r \in \mathbb{R}$, $0 < r < 1$, $\alpha \in \mathbb{R}$, $\alpha > 0$ and $\alpha \leq \frac{\bar{u}_k - \bar{u}_{k+1}}{2}$ with $\bar{u}_k$ and $\bar{u}_{k+1}$ denoting the kth largest and the $(k+1)$th largest element in u the function $g(\cdot)$ and $\text{sig}^r(\cdot)$ are defined in (3.4) and (3.10), respectively. In this neural network, the lower and upper bound are $l_i = 0$ and $h_i = 1$ for all i in (3.4) and $\text{sig}^r(\cdot)$ reduces to a scalar function with mapping $\mathbb{R} \to \mathbb{R}$.

3.5.2 Theoretical Results for Solving k-WTA with the Proposed Neural Network

On the stability, finite-time convergence and solution optimality of the neural network (3.43), we have conclusions stated in the following theorems.

Theorem 3.3 *The recurrent neural network (3.43) with $0 < r < 1$, $\rho > 0$ and $\varepsilon > 0$ is stable in the sense of Lyapunov and converges to an equilibrium point λ^* in finite time.*

Proof To prove the conclusion, we construct the following Lyapunov function,

$$V = \frac{|\mathbf{1}^T g(\frac{\rho u}{2\alpha} - \lambda \mathbf{1}) - \rho k|^{r+1}}{r+1} \tag{3.44}$$

where $|x|$ denotes the absolute value of a scalar x. The time derivative of V along the neural network trajectory (3.43) can be obtained as follows:

$$\dot{V} = -\dot{\lambda} \mathbf{1}^T D^+ g \mathbf{1} \text{sig}^r \left(\mathbf{1}^T g(\frac{\rho u}{2\alpha} - \lambda \mathbf{1}) - \rho k \right) \tag{3.45}$$

where $D^+ g$ denotes the upper right dini-derivative of the function $g(\frac{\rho \mu}{2\alpha} - \lambda \mathbf{1})$. According to the definition of $g(\cdot)$ in (3.4), we know $D^+ g$ is a diagonal matrix of the form $D^+ g = \text{diag}(J_1, J_2, \ldots, J_n)$ and the ith diagonal element J_i is as follows,

$$J_i = \begin{cases} 1 & \text{if } 0 \leq \dfrac{\rho u_i}{2\alpha} - \lambda < \rho \\[2mm] 0 & \text{if } \dfrac{\rho u_i}{2\alpha} - \lambda < 0 \text{ or } \dfrac{\rho u_i}{2\alpha} - \lambda \geq \rho \end{cases} \tag{3.46}$$

Further, we get

$$\begin{aligned} \dot{V} &= -\frac{\sum_{i=1}^n J_i}{\varepsilon} \left(\text{sig}^r \left(\mathbf{1}^T g(\frac{\rho u}{2\alpha} - \lambda \mathbf{1}) - \rho k \right) \right)^{2r} \\ &= -\frac{\sum_{i=1}^n J_i}{\varepsilon} \left(\mathbf{1}^T g(\frac{\rho u}{2\alpha} - \lambda \mathbf{1}) - \rho k \right)^{2r} \end{aligned} \tag{3.47}$$

Since $J_i \geq 0$ for all $i = 1, 2, \ldots, n$, $\sum_{i=1}^n J_i \geq 0$ and $\dot{V} \leq 0$, which means the neural network (3.43) is stable in the sense of Lyapunov.

To study the finite time convergence property, we first define a set $\mathbb{U} = \{\lambda \in \mathbb{R} \mid \dot{V} = 0\}$. Obviously, both the solution of $\mathbf{1}^T g(\frac{\rho u}{2\alpha} - \lambda\mathbf{1}) - \rho k = 0$ and the solution of $J_i = 0$ for all $i = 1, 2, \ldots, n$ are elements in $\mathbb{U}$. However, only the former one is invariant with the neural network trajectory (3.43). The latter one cannot stay on the neural network trajectory forever. According to LaSalle's invariance principle, we conclude that the neural network (3.43) asymptotically converges to the solution of $\mathbf{1}^T g(\frac{\rho u}{2\alpha} - \lambda\mathbf{1}) - \rho k = 0$, which means for any small ball centered at the solution of $\mathbf{1}^T g(\frac{\rho u}{2\alpha} - \lambda\mathbf{1}) - \rho k = 0$, we can always find a finite time t_1, after which λ stays in that ball forever. Moreover, since the solution of $J_i = 0$ for all $i = 1, 2, \ldots, n$ has no intersection with the solution of $\mathbf{1}^T g(\frac{\rho u}{2\alpha} - \lambda\mathbf{1}) - \rho k = 0$, we can always find a small enough ball centered at the solution of $\mathbf{1}^T g(\frac{\rho u}{2\alpha} - \lambda\mathbf{1}) - \rho k = 0$ but has no intersection with the solution of $J_i = 0$ for all $i = 1, 2, \ldots, n$ and also find the corresponding finite time t_1, after which λ stays inside the small enough ball forever. For λ in the above defined ball, $\sum_{i=1}^n J_i > 0$ (otherwise, the ball has intersection with the solution of $J_i = 0$ for all $i = 1, 2, \ldots, n$). The least value for $\sum_{i=1}^n J_i$ with $\sum_{i=1}^n J_i \neq 0$ is 1 according to (3.46). Therefore, we have for $t \geq t_1$:

$$
\dot{V} \leq -\frac{1}{\varepsilon} \left(\mathbf{1}^T g(\frac{\rho u}{2\alpha} - \lambda\mathbf{1}) - \rho k \right)^{2r}
$$
$$
= -\frac{1}{\varepsilon} \left((r+1)V \right)^{\frac{2r}{r+1}} \tag{3.48}
$$

As the auxiliary system $\dot{K} = -\frac{1}{\varepsilon}((r+1)K)^{\frac{2r}{r+1}}$ with $0 < r < 1$ and $K(t_1) = V(t_1)$ converges to zero in finite time, say Δt, we know $0 \leq V \leq K$ for $t \geq t_1$ according to the Comparison Lemma and thus the neural network also converges before the time $t_1 + \Delta t$. Since both Δt and t_1 are finite, we conclude that the neural network converges in finite time. This completes the proof.

The following Theorem reveals the relation between the equilibrium point of the neural network and the optimal solution to the k-WTA problem (3.33) (equivalent to the quadratic programming problem (3.34)).

Theorem 3.4 *Let λ^* be an equilibrium point of the neural network (3.43), the output of this neural network at λ^*, which is $x^* = \frac{1}{\rho}g(\frac{\rho u}{2\alpha} - \lambda^*\mathbf{1})$, is the optimal solution to the k-WTA problem (3.33).*

Proof The proof is similar to the proof of Theorem 3.2 and therefore omitted.

Remark 3.4 Theorem 3.4 reveals that the proposed neural network converges to the unique optimal solution of the k-WTA problem. Note that the equilibrium point λ^* is not necessary to be unique but all feasible λ^* lead to a unique x^* guaranteeing the optimality. The non-uniqueness of λ^* can be observed in Sect. 3.5.3.1.

3.5.3 k-WTA Simulations

In this section, we use two simulation examples to demonstrate the effectiveness of
the proposed method for solving the k-WTA problem.

3.5.3.1 Static k-WTA Problem

For the static k-WTA problem, we consider time invariant signals as the input. In
the simulation, we consider a k-WTA problem with $k = 5$ and $n = 15$ randomly
generated values as the input, which is $u = [1.2852, -1.3708, -3.2431, 4.4145,$
$-9.7269, -2.5188, 8.4537, 0.9296, -0.5223, -0.0693, -3.8206, 9.0168, 9.6399,$
$0.2713, 9.8518]$. The neural network parameters are chosen to be $\alpha = 0.1, \varepsilon = 10^{-6}$,
$r = 0.5$ and $\rho = 1$, respectively. Figure 3.11 illustrates the state trajectory under
different initializations. From this figure, we can see that the state λ converges to
different values under different initialization, which verifies the statement that the
equilibrium point λ^* is not unique as pointed out in Remark 3.4. For all these values
of λ^*, the outputs of the neural network are all equal to $[0, 0, 0, 1, 0, 0, 1, 0, 0, 0, 0,$
$1, 1, 0, 1]$, which means that u_4, u_7, u_{12}, u_{13} and u_{15} are the 5 largest ones in u.

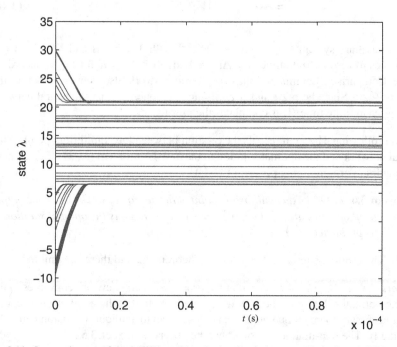

Fig. 3.11 State trajectory of the k-WTA problem in the simulation example in Sect. 3.5.3.1

3.5.3.2 Dynamic k-WTA Problem

For the dynamic k-WTA problem, the input signals are time varying and we use this example to test the real-time performance of our method. In the simulation, there are four input signals, which are $u_i = 10\sin(200\pi t + 0.4\pi(i - 1))$ for $i = 1, 2, 3, 4$, respectively. The four input signals compete for two winners, which means we choose $k = 2$. In addition, we choose $\alpha = 0.1$, $\rho = 1$, $\varepsilon = 10^{-6}$, $r = 0.5$ for the neural network. The four input signals and the output of the neural network are plotted in Fig. 3.12. From this figure, we can see the proposed neural network can successfully find the two largest signals in real time.

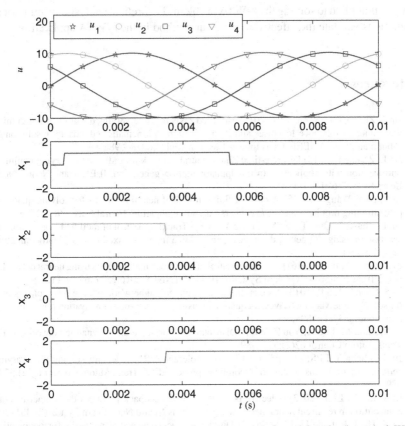

Fig. 3.12 Inputs and outputs of the proposed neural network for solving the dynamic k-WTA problem in Sect. 3.5.3.2

3.6 Summary

In this chapter, a finite-time dual neural network is presented to solve quadratic programming problems. Compared with existing recurrent neural networks, such as the dual neural network, the improved dual neural network, the simplified dual neural network, etc., which never converge in finite time, the proposed neural network has a finite-time converge property. Conditions for the stability, finite-time convergence and solution optimality to the quadratic programming problem are established in theory and simulations are performed to validate the finite-time convergence property and compare the convergence speed, sensitivity to additive noise and robustness against time delay under different choices of parameters. The finite-time dual neural network is applied to solving the k-WTA problem. Theoretical analysis and numerical simulations validate the effectiveness of our method in the k-WTA application.

References

1. Jin L, Zhang Y, Li S, Zhang Y (2016) Modified ZNN for time-varying quadratic programming with inherent tolerance to noises and its application to kinematic redundancy resolution of robot manipulators. IEEE Trans Industr Electron 63(11):6978–6988
2. Jin L, Zhang Y (2015) Discrete-time Zhang neural network for online time-varying nonlinear optimization with application to manipulator motion generation. IEEE Trans Neural Netw Learn Syst 27(6):1525–1531
3. Li S, Li Y, Wang Z (2013) A class of finite-time dual neural networks for solving quadratic programming problems and its k-winners-take-all application. Neural Netw 39(1):27–39
4. Jin L, Zhang Y, Qiao T, Tan M, Zhang Y (2016) Tracking control of modified Lorenz nonlinear system using ZG neural dynamics with additive input or mixed inputs. Neurocomputing 196(1):82–94
5. Li S, Zhang Y, Jin L (2016) Kinematic control of redundant manipulators using neural networks. IEEE Trans Neural Netw Learn Syst. doi:10.1109/TNNLS.2016.2574363 (In Press)
6. Li S, Chen S, Liu B (2013) Accelerating a recurrent neural network to finite-time convergence for solving time-varying Sylvester equation by using a sign-bi-power activation function. Neurocomputing 37(1):189–205
7. Li S, Zhou M, Luo X, You Z (2017) Distributed winner-take-all in dynamic networks. IEEE Trans Autom Control 62(2):577–589
8. Jin L, Zhang Y, Li S, Zhang Y (2017) Noise-tolerant ZNN models for solving time-varying zero-finding problems: A control-theoretic approach. IEEE Trans Autom Control 62(2):577–589
9. Li S, Liu B, Li Y (2013) Selective positive-negative feedback produces the winner-take-all competition in recurrent neural networks. IEEE Trans Neural Netw Learn Syst 24(2):301–309
10. Li S, He J, Rafique U, Li Y (2017) Distributed recurrent neural networks for cooperative control of manipulators: A game-theoretic perspective. IEEE Trans Neural Netw Learn Syst 28(2):415–426
11. Jin L, Zhang Y, Li S (2016) Integration-enhanced Zhang neural network for real-time varying matrix inversion in the presence of various kinds of noises. IEEE Trans Neural Netw Learn Syst 27(12):2615–2627
12. Li S, Li Y (2014) Nonlinearly activated neural network for solving time-varying complex sylvester equation. IEEE Trans Cybern 44(8):1397–1407

13. Jin L, Li S, La H, Luo X (2017) Manipulability optimization of redundant manipulators using dynamic neural networks. IEEE Trans Ind Electron pp(99):1–10. doi:10.1109/TIE.2017. 2674624 (In Press)

14. Zhang Y, Li S (2017) Predictive suboptimal consensus of multiagent systems with nonlinear dynamics. IEEE Trans Syst Man Cybern Syst pp(99):1–11. doi:10.1109/TSMC.2017.2668440 (In Press)

15. Li S, You Z, Guo H, Luo X, Zhao Z (2016) Inverse-free extreme learning machine with optimal information updating. IEEE Trans Cybern 46(5):1229–1241

16. Khan M, Li S, Wang Q, Shao Z (2016) CPS oriented control design for networked surveillance robots with multiple physical constraints. IEEE Trans Comput-Aided Des Integr Circuits Syst 35(5):778–791

17. Khan M, Li S, Wang Q, Shao Z (2016) Formation control and tracking for co-operative robots with non-holonomic constraints. J Intell Robot Syst 82(1):163–174

18. Li S, Cui H, Li Y (2013) Decentralized control of collaborative redundant manipulators with partial command coverage via locally connected recurrent neural networks. Neural Comput Appl 23(1):1051–1060

19. Hopfield JJ (1984) Neurons with graded response have collective computational properties like those of two-state neurons. Proc Natl Acad Sci 81(10):3088–3092

20. Zhang S, Constantinides AG (1992) Lagrange programming neural networks. IEEE Trans Circuits Syst II: Analog Digital Signal Process 39(7):441–452

21. Wu X, Xia Y, Li J, Chen W (1996) A high-performance neural network for solving linear and quadratic programming problems. IEEE Trans Neural Netw 7(3):643–651

22. Hu X, Zhang B (2009) A new recurrent neural network for solving convex quadratic programming problems with an application to the k-winners-take-all problem. IEEE Trans Neural Netw 20(4):654–664

23. Liu S, Wang J (2006) A simplified dual neural network for quadratic programming with its k-wta application. IEEE Trans Neural Netw 17(6):1500–1510

24. Jin L, Zhang Y (2016) Continuous and discrete Zhang dynamics for real-time varying nonlinear optimization. Numer Algorithm 73(1):115–140

25. Jin L, Zhang Y, Qiu B (2016) Neural network-based discrete-time Z-type model of high accuracy in noisy environments for solving dynamic system of linear equations. Neural Comput Appl. doi:10.1007/s00521-016-2640-x (In Press)

26. Jin L, Zhang Y (2015) G2-type SRMPC scheme for synchronous manipulation of two redundant robot arms. IEEE Trans Cybern 45(2):153–164

27. Wang J (2010) Analysis and design of a k-winners-take-all model with a single state variable and the heaviside step activation function. IEEE Trans Neural Netw 21(9):1496–1506

28. Liu Q, Wang J (2011) Finite-time convergent recurrent neural network with a hard-limiting activation function for constrained optimization with piecewise-linear objective functions. IEEE Trans Neural Netw 22(4):601–613

29. Bhat S, Bernstein D (2000) Finite-time stability of continuous autonomous systems. SIAM J Control Optim 38(1):751–766

30. Boyd S, Vandenberghe L (2004) Convex optimization. Cambridge University Press

31. Jin L, Li S (2017) Distributed task allocation of multiple robots: A control perspective. IEEE Trans Syst Man Cybern Syst pp(99):1–9

Chapter 4
Competition Based on Selective Positive-Negative Feedback

Abstract In this chapter, we make steps in that direction and present a simple model, which produces the winner-take-all competition by taking advantage of selective positive-negative feedback through the interaction of neurons via p-norm. Compared to models presented in Chaps. 1, 2 and 3, this model has an explicit explanation of the competition mechanism. The ultimate convergence behavior of this model is proven analytically. The convergence rate is also discussed. Simulations are conducted in the static competition and the dynamic competition scenarios. Both theoretical and numerical results validate the effectiveness of the dynamic equation in describing the nonlinear phenomena of winner-take-all competition.

Keywords Winner-take-all competition · Recurrent neural networks · Selective positive-negative feedback · Global stability · Numerical simulations

4.1 Introduction

Winner-take-all refers to the phenomena that agents in a group compete with each others for activation and only the one with the highest input stays active while all the others deactivated. It widely exists in nature and the society: for most plants, the main central stem, which only appears slightly stronger than other side stems at the very beginning of the plant development, grows more and more strongly and eventually is dominant over others [1]. It has been observed in the society that once a firm gets ahead, it is more likely to become better and better over time while the others will fall further behind [2]. Neuroscientists find that the contrast gain in the visual systems comes from a winner-take-all competition among overlapping neurons [3]. Other examples of the winner-take-all competition include decision making in the cortex [4], animal behaviors [5], competition-based motion generation of multiple redundant manipulators [6], etc.

Apart from the natures of distributed-storage and high-speed parallel-processing, neural networks can be readily implemented by hardware and thus have been widely applied in various fields, including the competition phenomena, [7–24]. The N species Lotka–Volterra model is often used to model the competitive interaction

© The Author(s) 2018
S. Li and L. Jin, *Competition-Based Neural Networks with Robotic Applications*,
SpringerBriefs in Applied Sciences and Technology,
DOI 10.1007/978-981-10-4947-7_4

between species. Under elaborately selected parameters, the N species Lotka–Volterra model is able to demonstrate the winner-take-all competition. This model is applied in [25] to generate the winner-take-all competition. In the field of computational neuroscience, the FitzHugh–Nagumo model is often used to describe the dynamic interaction of neurons. In some situations, the neurons interact with each other in a winner-take-all manner and the winner spikes. Inspired by this fact, this describing model is in turn applied to generate the winner-take-all behavior [26, 27]. In [26], the authors show that the model outputs oscillations under a set of parameter setups and the oscillatory amplitude of the winner is greater than the spiking threshold while the amplitude of the losers are much less than threshold. In [27], theoretical analysis on the stability and convergence of a large scale winner-take-all networks is conducted by using nonlinear contraction theory. In addition, the authors show that the presented network is stable for a range of parameters. In [28–31], the winner-take-all problem is solved by modeling the problem as an optimization problem. In [28], a combinatorial optimization solver is presented to solved the problem. In [29], the problem is model as a convex quadratic programming problem and a recurrent neural network developed for solving constrained quadratic programming is applied to solve it. Following the same problem formulation, the neural network presented in [29] is simplified in [30] by tailoring the structure and taking advantage of the nonlinearity provided by a saturation function used in the model. In [31], an one-layer recurrent neural network is developed to solve the winner-take-all competition by modeling the problem as a constrained linear programming. Although the optimization based approach solves the problem accurately, operations such as saturation function, matrix multiplication of the state vector, etc., are often necessary in the iterations to approach the desired solution and thus are often computationally intensive. In addition, the resulting dynamics are often complicated and are often difficult to explain the winner-take-all mechanism from its dynamic equations.

Although many models have been presented to explain and generate the winner-take-all behavior [25–31], these models are often very complicated due to the compromise with experimental realities in the particular fields. Consequently, the essence of the winner-take-all competition may be embedded in the interaction dynamics of those models, but difficult to tell from the sophisticated dynamic equations. Motivated by this, we develop a simple neural network model to solve the problem. The presented model has a star communication topology between neurons and is scalable to situation with a large number of competitors. The model is described by an ordinary equation with the space dimension equal to the number of competitors. In addition, compared with the model using the Euclidean norm for global information exchange, the presented model extend the results to the more general p-norm cases. Moreover, the presented model demonstrates different robustness and convergence speed for different choice of parameters and thus allow the user to choose a set of parameters for better performance in applications.

4.2 Preliminaries

In this section, we present some useful preliminaries for p-norm and system stability. We first present preliminaries on p-norm.

For a n-dimensional vector $x = [x_1, x_2, \ldots, x_n]^T$ with $x_i \in \mathbb{R}$ for $i = 1, 2, \ldots, n$, its p-norm, denoted as $\|x\|_p$, is defined as follows,

$$\|x\|_p = (|x_1|^p + |x_2|^p + \cdots + |x_n|^p)^{\frac{1}{p}}, \tag{4.1}$$

where $p > 0$.

For $\|x\|_p$, the following partial derivative results hold,

$$
\begin{aligned}
\frac{\partial \|x\|_p^p}{\partial x_i} &= \frac{\partial(|x_1|^p + |x_2|^p + \cdots + |x_n|^p)}{\partial x_i} \\
&= \frac{\partial |x_i|^p}{\partial x_i} = \frac{\partial |x_i|^p}{\partial x_i} = p|x_i|^{p-1}\text{sgn}(x_i),
\end{aligned} \tag{4.2}
$$

where $\text{sgn}(\cdot)$ is the sign function defined as,

$$
\text{sgn}(u) = \begin{cases} 1 & \text{if } u > 0 \\ 0 & \text{if } u = 0 \\ -1 & \text{if } u < 0, \end{cases} \tag{4.3}
$$

with $u \in \mathbb{R}$.

Based on the partial derivative of $\|x\|_p$ shown in (4.2), the gradient of $\frac{1}{p}\|x\|_p^p$ can be obtained as,

$$\nabla \frac{1}{p}\|x\|_p^p = \text{sig}^{p-1}(x), \tag{4.4}$$

for $p > 0$ with the operator 'sig$^k(\cdot)$' defined as,

$$\text{sig}^k(x) = [|x_1|^k \text{sgn}(x_1), |x_2|^k \text{sgn}(x_2), \ldots, |x_n|^k \text{sgn}(x_n)]^T, \tag{4.5}$$

where $x = [x_1, x_2, \ldots, x_n]^T$. According to this definition, we can directly obtain the following equation,

$$x^T \text{sig}^k(x) = \|x\|_{k+1}^{k+1}. \tag{4.6}$$

For following inequalities hold for the estimation of p-norms in different p values.

$$\|x\|_p \leq \|x\|_r \leq n^{\frac{1}{r}-\frac{1}{p}}\|x\|_p, \tag{4.7}$$

where $p > r > 0$, n represents the dimension of the vector x.

The following results will be used later as tools for convergence analysis.

Definition 4.1 ([32]) A continuous function $\alpha : [0, a) \to [0, \infty)$ is said to belong to class $\mathscr{K}$ if it is strictly increasing and $\alpha(0) = 0$. It is said to belong to class $\mathscr{K}_\infty$ if $a = \infty$ and $\alpha(r) \to \infty$ as $r \to \infty$.

Lemma 4.1 ([32]) *Let* $\mathbb{D} \subset \mathbb{R}^n$ *be a domain that contains the origin and* $V : [0, \infty) \times \mathbb{D} \to \mathbb{R}$ *be a continuous differentiable function such that*

$$\alpha_1(\|x\|) \le V(t, x) \le \alpha_2(\|x\|), \tag{4.8}$$

$$\dot{V} = \frac{\partial V}{\partial t} + \frac{\partial V}{\partial x} f(t, x) \le -W(x), \quad \forall \|x\| \ge \mu > 0, \tag{4.9}$$

$\forall t \ge 0$ *and* $\forall x \in \mathbb{D}$, *where* α_1 *and* α_2 *are class* $\mathscr{K}$ *functions and* $W(x)$ *is a continuous positive definite function. Take* $r > 0$ *such that* $\mathbb{B}_r \subset \mathbb{D}$ *and suppose that* $\mu < \alpha_2^{-1}(\alpha_1(r))$. *Then, for every initial state* $x(t_0)$, *satisfying* $\|x(t_0)\| \le \alpha_2^{-1}(\alpha_1(r))$, *there is* $T \ge 0$ *(dependent on* $x(t_0)$ *and* μ*) such that the solution of* $\dot{x} = f(t, x)$ *satisfies,*

$$\|x(t)\| \le \alpha_1^{-1}(\alpha_2(\mu)) \quad \forall t \ge t_0 + T. \tag{4.10}$$

Moreover, if $\mathbb{D} = \mathbb{R}^n$ *and* α_1 *belongs to class* $\mathscr{K}_\infty$, *then the result (4.15) holds for any initial state* $x(t_0)$, *with no restriction on how large* μ *is.*

The following Lemma is also useful for the analysis of the ultimate behavior of a dynamic system,

Lemma 4.2 ([33]) *Let* $\Omega \subset \mathbb{D}$ *be a compact set that is positively invariant with respect to* $\dot{x} = f(x)$. *Let* $V : \mathbb{D} \to \mathbb{R}$ *be a* C^1-*function such that* $\dot{V}(x) \le 0$ *on* Ω. *Let* $\mathbb{E}$ *be the set of all points in* Ω *such that* $\dot{V}(x) = 0$. *Let* $\mathbb{M}$ *be the largest invariant set in* $\mathbb{E}$. *Then, every solution starting in* Ω *approaches* $\mathbb{M}$ *as* $t \to \infty$.

The mapping V in Lemma 4.2 is not necessary to be positive definite, which is a major difference from the Lyapunov function in conventional stability analysis of dynamic systems [33]. Instead, V is required to be be a continuous differentiable function in Lemma 4.2, which is much looser than the positive definite requirement.

4.3 The Winner-Take-All Neural Network

4.3.1 The Neural Network Based Winner-Take-All Problem

In this chapter, similar to the previous chapters, we are concerned with a neural network based approach to find the winner in a group of competitors. Concretely, we want to find $i^* = \text{argmax}\{u_1, u_2, \ldots, u_n\}$ for the input vector $u = [u_1, u_2, \ldots, u_n]^T \in \mathbb{R}^n$ with $u_i \in \mathbb{R}$ by using neural networks, i.e., to find the winner among u_1, u_2, ..., u_n by neural networks.

4.3.2 Neuro-Dynamics

Inspired by the normalized recurrent neural network [34] and the use of the general norm on modeling the power of signals [35], we presente a recurrent neural network with a general p-norm as the regulation term for winner-take-all competition. The presented model has the following dynamic for the ith neuron in a group of totally n neurons,

$$\dot{x}_i = c_0(u_i - c_1\|x\|_{p+1}^{p+1})|x_i|^p\mathrm{sgn}(x_i), \tag{4.11}$$

where $x_i \in \mathbb{R}$ denotes the state of the i neuron, $u_i \in \mathbb{R}$ is the input and $u_i \geq 0$, $u_i \neq u_j$ for $i \neq j$, $p \in \mathbb{R}$, $p \geq 0$, $\|x\|_{p+1}$ is the $(p+1)$-norm of the state vector $x = [x_1, x_2, \ldots, x_n]^T$, $c_0 \in \mathbb{R}$, $c_0 > 0$ and $c_1 \in \mathbb{R}$, $c_1 > 0$ are both constant.

The dynamic Eq. (4.11) can be written into the following compact form by stacking up the state for all neurons,

$$\dot{x} = c_0\big(u \circ \mathrm{sig}^p(x) - c_1\|x\|_{p+1}^{p+1}\mathrm{sig}^p(x)\big), \tag{4.12}$$

where $x = [x_1, x_2, \ldots, x_n]^T$, $u = [u_1, u_2, \ldots, u_n]^T$, the operator 'o' represents the multiplication in component-wise, i.e., $u \circ x = [u_1x_1, u_2x_2, \ldots, u_nx_n]^T$.

Remark 4.1 The presented neural network can be regarded as a black box. The ith neuron in the network receives input u_i and outputs x_i through the dynamic interactions with other neurons. As will be proved in Sect. 4.4, with the presented model (4.11), the winner neuron $i^* = \mathrm{argmax}\{u_1, u_2, \ldots, u_n\}$ can be identified by checking whether $\lim_{t\to\infty} x_i(t) = 0$ (if $\lim_{t\to\infty} x_i(t) \neq 0$, $i = i^*$ and otherwise, $i \neq i^*$).

Remark 4.2 The neuro-dynamics described by (4.11) is connected in a star topology. As can be observed from (4.11), the ith neuron is only connected to the central node, which computes the p-norm of the whole network state values. The information exchange between neurons comes indirectly from their direct interaction with the central node.

Remark 4.3 Choosing the Euclidean norm, which corresponds to the special case of (4.11) by choosing $p = 1$, the presented model reduces to the following in vector form,

$$\dot{x} = c_0(u \circ x - c_1\|x\|^2 x), \tag{4.13}$$

where $\|\cdot\|$ represents the Euclidean norm. In other words, the presented model is a generalization from (4.13), which uses the Euclidean norm to the general p-norm scheme. Note that this generalization is not trivial as the p-norm function $y = \|x\|_p$ corresponds to different level sets (see Fig. 4.1) and thus lead to completely different dynamic evolution of x in (4.11).

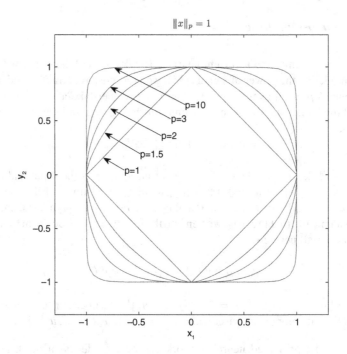

Fig. 4.1 The level set for $\|x\|_p = 1$ in two-dimensional space with different value of p

Remark 4.4 Particularly for $p = 0$, the presented model (4.11) reduces to the following in vector form,

$$\dot{x} = c_0\big(u \circ \mathrm{sgn}(x) - c_1\|x\|_1 \mathrm{sgn}(x)\big), \tag{4.14}$$

where $\mathrm{sgn}(x) = [\mathrm{sgn}(x_1), \mathrm{sgn}(x_1), \ldots, \mathrm{sgn}(x_n)]^T$ for $x = [x_1, x_2, \ldots, x_n]^T$ with $\mathrm{sgn}(\cdot)$ being the sign for scalar entries. Note that this is a typical recurrent neural network with hard-limiting activation function and is often able to demonstrate a finite-time convergence as shown in [36]. In addition, it is noteworthy that the global information term $\|x\|_p^p$ reduces to $\|x\|_1$ in (4.14), which is the norm widely used in machine learning for data sparsification due to its ability to approximate the cardinality $\|x\|_0$. It will be an interesting topic to investigate the protocol from the perspective of sparse optimization.

4.4   Convergence Results

In this section, theoretical results on the dynamic system (4.11) are presented. The rigorous proof of the main results needs the uses of LaSalle's invariant set principle [33], local stability analysis and the ultimate boundedness theory [32].

With Lemma 4.1, we are able to prove the following lemma for our main result,

Lemma 4.3 *There exists $T \geq 0$ (dependent on $x(t_0)$ and μ) such that the solution of the neuron dynamic Eq. (4.12) satisfies,*

$$\|x(t)\| \leq \mu \quad \forall t \geq t_0 + T, \tag{4.15}$$

where $\mu = \mu_0 (\frac{u_{max}+\mu_1}{c_1})^{\frac{1}{p+1}}$ *with* $\mu_1 > 0$ *being any positive constant,* $\mu_0 = \max\{n^{(\frac{1}{2}-\frac{1}{p+1})}, 1\}$ *and* $u_{max} = \max\{u_1, u_2, \ldots, u_n\}$.

Proof We prove the result by following the framework of Lemma 4.1. Let $\mathbb{D} = \mathbb{R}^n$, $V = \frac{1}{2}x^T x$ and $\alpha_1(\|x\|) = \alpha_2(\|x\|) = \frac{1}{2}\|x\|^2 = V$. For V, we have

$$
\begin{aligned}
\dot{V} &= x^T \dot{x}, \\
&= c_0 x^T \left(u \circ \text{sig}^p(x) - c_1 \|x\|_{p+1}^{p+1} \text{sig}^p(x) \right), \\
&= c_0 x^T \left(diag(u)\text{sig}^p(x) - c_1 \|x\|_{p+1}^{p+1} \text{sig}^p(x) \right), \\
&\leq c_0 (u_{max} - c_1 \|x\|_{p+1}^{p+1}) x^T \text{sig}^p(x), \\
&= c_0 (u_{max} - c_1 \|x\|_{p+1}^{p+1}) \|x\|_{p+1}^{p+1}.
\end{aligned} \tag{4.16}
$$

Note that $diag(u)$ is a diagonal matrix and its largest eigenvalue is u_{max}, from which the last inequality in (4.16) is obtained. According to the inequalities given in (4.7), we have

$$
\|x\| \leq \begin{cases} n^{(\frac{1}{2}-\frac{1}{p+1})}\|x\|_{p+1} & \text{for } p > 1, \\ \|x\|_{p+1} & \text{for } p \leq 1, \end{cases}
$$

where n represents the dimension of x. Inequality (4.17) is equivalent to the following,

$$\|x\| \leq \mu_0 \|x\|_{p+1}, \tag{4.17}$$

where $\mu_0 = \max\{n^{(\frac{1}{2}-\frac{1}{p+1})}, 1\}$. Together with (4.16), we have,

$$\dot{V} \leq c_0 \left(u_{max} - c_1 \left(\frac{\|x\|}{\mu_0} \right)^{p+1} \right) \|x\|_{p+1}^{p+1}, \tag{4.18}$$

With this result, it is sufficient for $\left(u_{max} - c_1 \left(\frac{\|x\|}{\mu_0} \right)^{p+1} \right) \leq -\mu_1 \leq 0$, i.e., $\|x\| \geq \mu_0 \left(\frac{u_{max}+\mu_1}{c_1} \right)^{\frac{1}{p+1}}$, to guarantee the following inequality,

$$\dot{V} \leq -c_0 \mu_1 \|x\|_{p+1}^{p+1}, \tag{4.19}$$

with $\mu_1 > 0$. This results fall into the framework of Lemma 4.1 by choosing $\mu = \mu_0 \left(\frac{u_{max} + \mu_1}{c_1} \right)^{\frac{1}{p+1}}$ and $W(x) = c_0 \mu_1 \|x\|_{p+1}^{p+1}$ in (4.9). Therefore, we conclude, according to Lemma 4.1, that

$$\|x\| \leq \alpha_1^{-1}(\alpha_2(\mu)) = \mu = \mu_0 \left(\frac{u_{max} + \mu_1}{c_1} \right)^{\frac{1}{p+1}} \quad \forall t > t_0 + T, \tag{4.20}$$

for any initialization of x. This completes the proof.

This lemma reveals the state of the dynamic model (4.12) is ultimately bounded inside a compact super ball in $\mathbb{R}^n$ with radius μ. In other words, this super ball is positively invariant with respect the system dynamic (4.12). With this result on hand, we can confine our analysis in this super ball for further investigation of the system behaviors by applying LaSalle's invariant set principle.

Theorem 4.1 *The solution of the system involving n dynamic neurons with the ith neuron described by (4.11) globally approaches 0 for $i \neq i^*$ and approaches $\left(\frac{u_{i^*}}{c_1} \right)^{\frac{1}{p+1}} e_{i^*}$ (or $-\left(\frac{u_{i^*}}{c_1} \right)^{\frac{1}{p+1}} e_{i^*}$) for $i = i^*$ as $t \to \infty$, provided any initialization with the initial value of the i^* neuron positive (or negative), where i^* denotes the label of the winner, i.e., $i^* = argmax\{u_1, u_2, \ldots, u_n\}$.*

Proof There are three steps for the proof. The first step is to prove that the state variable ultimately converges to a set consisting of a limit number of points, the second step proves there are only two single point among the candidates is stable and the third step gives the condition on initial conditions to decide which stable equilibrium point the system will converge to.

Step 1: According to Lemma 4.3, the state variable x in the system dynamic (4.12) is ultimately bounded by a compact super ball in $\mathbb{R}^n$ with radius μ implying this super ball is positively invariant with respect the system dynamic (4.12) and the super ball $\{x \in \mathbb{R}^n | \|x\| \leq \mu\}$ is qualified to be the set Ω in Lemma 4.2.

Let $V = V_1 + V_2$, with

$$V_1 = -\frac{1}{p+1} \sum_{i=1}^{n} u_i |x_i|^{p+1}, \tag{4.21}$$

$$V_2 = \frac{c_1}{2(p+1)} \|x\|_{p+1}^{2p+2}. \tag{4.22}$$

Apparently, V is a C^1-function. For V_1, we have,

$$\dot{V}_1 = -\sum_{i=1}^{n} u_i |x_i|^p \text{sgn}(x_i) \dot{x}_i = -\left(\text{sig}^p(x) \right)^T \text{diag}(u) \dot{x}. \tag{4.23}$$

For V_2, we have,

$$
\begin{aligned}
\dot{V}_2 &= \frac{c_1}{2(p+1)} \frac{d(\|x\|_{p+1}^{p+1})^2}{dt} = \frac{c_1}{(p+1)} \|x\|_{p+1}^{p+1} \frac{d(\|x\|_{p+1}^{p+1})}{dt} \\
&= \frac{c_1}{(p+1)} \|x\|_{p+1}^{p+1} (\nabla \|x\|_{p+1}^{p+1})^T \dot{x} = \frac{c_1}{(p+1)} \|x\|_{p+1}^{p+1} (p+1) (\text{sig}^p(x))^T \dot{x} \\
&= c_1 \|x\|_{p+1}^{p+1} (\text{sig}^p(x))^T \dot{x}.
\end{aligned}
\tag{4.24}
$$

Accordingly,

$$
\begin{aligned}
\dot{V} &= \dot{V}_1 + \dot{V}_2 \\
&= -\left((\text{sig}^p(x))^T \text{diag}(u) - c_1 \|x\|_{p+1}^{p+1} (\text{sig}^p(x))^T \right) \dot{x} \\
&= -c_0 \| u \circ \text{sig}^p(x) - c_1 \|x\|_{p+1}^{p+1} \text{sig}^p(x) \|^2 \\
&\leq 0.
\end{aligned}
\tag{4.25}
$$

According to the expression of $\dot{V}$ obtained in (4.25), we find $\text{diag}(u)\text{sig}^p(x) = c_1 \|x\|_{p+1}^{p+1}\text{sig}^p(x)$ by letting $\dot{V} = 0$. Note that $\text{diag}(u)\text{sig}^p(x) = c_1 \|x\|_{p+1}^{p+1}\text{sig}^p(x)$ is an eigenvector equation relative to the matrix $\text{diag}(u)$ and the vector $\text{sig}^p(x)$. Note that the eigenvalue and eigenvector pairs of the diagonal matrix $\text{diag}(u)$ are u_i and ke_i for $i = 1, 2, \ldots n$, with $k \in \mathbb{R}$ is a scaling constant and e_i denoting a n-dimensional vector with the ith component 1 and all the other component 0. Therefore, the solution for $\text{diag}(u)\text{sig}^p(x) = c_1\|x\|_{p+1}^{p+1}\text{sig}^p(x)$ is the solution of the two equations $c_1\|x\|_{p+1}^{p+1} = u_i$ and $\text{sig}^p(x) = ke_i$ for $i = 1, 2, \ldots, n$ (i.e., $x_e = \pm(\frac{u_i}{c_1})^{\frac{1}{p+1}} e_i$ by solving the two equations) and the trivial solution $x_e = 0$.

Define the set $\mathbb{M} = \{0, \pm(\frac{u_i}{c_1})^{\frac{1}{p+1}} e_i \text{ for } i = 1, 2, \ldots, n\}$. According to Lemma 4.2, every solution starting in $\Omega = \{x \in \mathbb{R}^n | \|x\| \leq \mu\}$ approaches $\mathbb{M}$ as $t \to \infty$. Together with the fact proven in Lemma 4.3 that every solution stays in Ω ultimately, we conclude that every solution with any initialization approaches $\mathbb{M}$ as $t \to \infty$.

Step 2: The first step in this proof shows there are several candidate fixed points to stay for the dynamic system. In this step, we show that all those fixed points in $\mathbb{M}$ are unstable except for the one corresponding to the winner, i.e., $x = \pm(\frac{u_{k*}}{c_1})^{\frac{1}{p+1}} e_{k*}$, where $k^* = \text{argmax}\{u_1, u_2, \ldots, u_n\}$. To show the instability of some equilibrium points, we only need to show that there exists a streamline starting from that equilibrium point to elsewhere, which is equivalent to the fact that there exists a streamline from a non-equilibrium point to the equilibrium point for the new dynamic system with time t replaced by $-t$ (note that for an autonomous system, replacing t with $-t$ meaning that the initial state of the original system is identical to the ultimate state of the new system). Following this idea, we consider the following auxiliary system with reversed time direction,

$$\dot{x}_i = -c_0(u_i - c_1\|x\|_{p+1}^{p+1})|x_i|^p\mathrm{sgn}(x_i), \qquad (4.26)$$

and we need to show that there exists a state x_0, the streamline of (4.26) starting from which ends up at the equilibrium point x_e.

For $x_e = 0$, we choose $x_0 = ke_1$, where $k > 0$ is a small positive constant and e_1 denotes a n-dimensional vector with the first component 1 and all the other component 0. Clearly, x_j for $j = 2, 3, \ldots, n$ starting from $x_0 = ke_1$ for the auxiliary system (4.26) stays at $x_j = 0$ in values since $\dot{x}_j = 0$ for them while $\dot{x}_1 = -c_0(u_1 - c_1\|x\|_{p+1}^{p+1})|x_1|^p\mathrm{sgn}(x_1) < 0$ for $x_1 > 0$ and small enough k, which means x_1 keeps reducing to zero. Therefore, we conclude that $x_e = 0$ is unstable.

For $x_e = (\frac{u_i}{c_1})^{\frac{1}{p+1}}e_i$ with $i \neq i^*$ (i^* denotes the winner neuron), we choose $x_0 = x_e + ke_{i*}$ with $k > 0$ being a constant to test the convergence. For $j \neq i^*$, the value of x_j of the auxiliary system (4.26) starting from $x_0 = x_e + ke_{i*}$ stays at $x_j = x_{ej}$ in values since $u_j|x_j|^p\mathrm{sgn}(x_j) = \|x\|_{p+1}^{p+1}|x_j|^p\mathrm{sgn}(x_j)$ (i.e., $\dot{x}_j = 0$). For $j = i^*$, $\dot{x}_{i*} = -c_0(u_{i*} - c_1\|x_e + ke_{i*}\|_{p+1}^{p+1})|k|^p\mathrm{sgn}(k)$ at $x_j = k$. Note that $x_e = (\frac{u_i}{c_1})^{\frac{1}{p+1}}e_i$ implies $c_1\|x_e + ke_{i*}\|_{p+1}^{p+1} = u_i < u_{i*}$. In addition, $\|x_e + ke_{i*}\|_{p+1}^{p+1} \approx \|x_e\|_{p+1}^{p+1}$ for small enough $k > 0$. Accordingly, $\dot{x}_{i*} < 0$ for small enough $k > 0$. Therefore, we conclude that $x_e = (\frac{u_i}{c_1})^{\frac{1}{p+1}}e_i$ is unstable.

It is worth noting that for $i = i^*$, $x_0 = x_e + ke_i$. Note that $x_e + ke_i = ((\frac{u_i}{c_1})^{\frac{1}{p+1}}+1)e_i$ and $\|x_e + ke_i\|_{p+1}^{p+1} = ((\frac{u_i}{c_1})^{\frac{1}{p+1}}+1)^{p+1} > ((\frac{u_i}{c_1})^{\frac{1}{p+1}})^{p+1} = \|x_e\|_{p+1}^{p+1}$ for any $k > 0$. Also computing $p + 1$ norm on both sides of $x_e = (\frac{u_i}{c_1})^{\frac{1}{p+1}}e_i$ generates $u_i = c_1\|x_e\|_{p+1}^{p+1}$. Together with $\|x_e + ke_i\|_{p+1}^{p+1} > \|x_e\|_{p+1}^{p+1}$, we get $u_i - c_1\|x_e + ke_i\|_{p+1}^{p+1} < 0$ for $k > 0$. Also note that x_{ei} dominates over k for small enough $k > 0$ in $\mathrm{sgn}(x_{ei}+k)$ and $|x_{ei}+k|$, we thus conclude $\dot{x}_i = -c_0(u_i - c_1\|x_e + ke_i\|_{p+1}^{p+1})|x_{ei} + k|^p\mathrm{sgn}(x_{ei} + k) > 0$ for small enough positive constant k (recall $x_{ei} > 0$), which is different from the cases with $i \neq i^*$.

The instability of $x_e = -(\frac{u_i}{c_1})^{\frac{1}{p+1}}e_i$ with $i \neq i^*$ (i^* denotes the winner neuron) can be similarly proved and thus omitted.

Step 3: Actually, we can conclude that the steady state value is $(\frac{u_{i*}}{c_1})^{\frac{1}{p+1}}e_{i*}$ if the initial state of the winner is positive while it is $-(\frac{u_{i*}}{c_1})^{\frac{1}{p+1}}e_{i*}$ if the initial value is negative by noting that $\dot{x}_{i*} = 0$ when $x_{i*} = 0$ in (4.11) for $i = i^*$, which means the state value x_{i*} will never cross the critical value $x_i^* = 0$.

In summary, we conclude that every solution approaches $x_e = (\frac{u_{i*}}{c_1})^{\frac{1}{p+1}}e_{i*}$ (or $x_e = -(\frac{u_{i*}}{c_1})^{\frac{1}{p+1}}e_{i*}$) ultimately, provided any initialization with the initial value of the i^* neuron positive (or negative), where $i^* = \mathrm{argmax}\{u_1, u_2, \ldots, u_n\}$ and e_{i*} being a n-dimensional vector with the i^*th component 1 and all the other component 0. Entrywisely, the solution approaches $x_i = 0$ for $i \neq i^*$ and $x_{i*} = \pm(\frac{u_{i*}}{c_1})^{\frac{1}{p+1}}$, which completes the proof.

4.5   Discussion on One-Sided Competition Versus Closely-Matched Competition

In this section, we provide an comparison of an one-sided competition and a closely-matched competition using the presented model.

Imagine a football game. There are only two neurons and the competition happens between the two teams. If one team has an overwhelming strength, it may demonstrate an obvious advantage over its opponent in a very early stage while it often takes a relatively long time for a fierce competition between two closely-matched teams to demonstrate a clear win and loss. Analogously, we may expect to observe a fast convergence in the winner-take-all competition, where a distinct advantage for one neuron over others exists while a slow convergence for closely-matched competitions. Theoretically speaking, this expectation corresponds to the statement that the convergence rate of the winner-take-all competition has a strong dependence on the comparisons of the input value of the winner and the input values of the others. This phenomena can be explained by the model (4.11). For simplicity, we consider the case with parameter $c_0 = c_1 = p = 1$ in (4.11), i.e., the following dynamic equations,

$$\dot{x} = u \circ x - \|x\|^2 x, \tag{4.27}$$

where $\|x\|$ represents the Euclidean norm of the vector x. As there exists a strong nonlinearity in (4.27), it is difficult to analyze the convergence rate directly. Nevertheless, we can approximately analyze the convergence rate by considering its linearization about the equilibrium point. According to Theorem 4.1, the stable equilibrium point is $x_e = \pm(\frac{u_{i^*}}{c_1})^{\frac{1}{p+1}} e_{i^*} = \pm\sqrt{u_{i^*}}e_{i^*}$, where i^* denotes the label of the winner and e_{i^*} denotes a n-dimensional vector with the i^*th element being 1 and all the other elements being zeros. The linearized system around this fixed point is as follows,

$$\dot{x} = \left(\text{diag}(u) - 2x_e x_e^T - \|x_e\|^2 I_n\right)x, \tag{4.28}$$

where I_n is a $n \times n$ identity matrix. The system matrix of the above system is a diagonal matrix and its jth diagonal component, which is also its jth eigenvalue, is $(u_j - u_{i^*})$ for $j \neq i^*$ and $-2u_{i^*}$ for $j = i^*$. The linear system (4.28) has all eigenvalues negative and its convergence rate is determined by the largest eigenvalue $2u_{i^*}$. In other words, the model (4.27) has an approximate convergence rate $2u_{i^*}$.

4.6   Simulation Examples

In this section, simulations are provided to illustrate the the winner-take-all competition phenomena generated by the neural dynamic (4.11). We consider two sceneries: one is static competition, i.e., the input u is constant and one is dynamic competition, i.e., the input u is time-varying.

4.6.1 Static Competition

4.6.1.1 Simulation Setup

For the static competition problem, we consider time invariant signals as the input. In the simulation, we consider a problem with $n = 15$ neurons. The input u is randomly generated between 0 and 1, which is $u = [0.9619, 0.0046, 0.7749, 0.8173, 0.8687, 0.0844, 0.3998, 0.2599, 0.8001, 0.4314, 0.9106, 0.1818, 0.2638, 0.1455, 0.1361]$, and the state is randomly initialized between -1 and 1, which is $x(0) = [0.7386, 0.1594, 0.0997, -0.7101, 0.7061, 0.2441, -0.2981, 0.0265, -0.1964, -0.8481, -0.5202, -0.7534, -0.6322, -0.5201, -0.1655]$. In the simulation, we choose $c_0 = c_1 = 1$.

4.6.1.2 Convergence

Figures 4.2, 4.3 and 4.4 show the evolution of state values of all neurons with time under different choice of the parameter p. From these figures, it can be observed that only a single state (corresponds to the 1st neuron, which has the largest value in u) has a non-zero value eventually and all the other state values are suppressed to zero. Also, the value of x_1 approaches $u_5^{\frac{1}{p+1}}$ (note that we choose $c_1 = 1$ in this simulation example), which is consistent with the claim made in Theorem 4.1 since the initial value $x_1(0) > 0$.

4.6.1.3 Convergence Speed

As can be observed in Figs. 4.2, 4.3 and 4.4, it takes about 14 s for the model to converge for $p = 0$, 30 s for $p = 0.5$, 80 s for $p = 1$, 350 s for $p = 1.5$ and more than 1000 s for $p = 2$, respectively, which implies that a faster convergence can be obtained by choosing a smaller p in the presented model for $p \geq 0$.

4.6.1.4 Robustness Against Additive Noise

In the real implementation of the presented model, additive noise, resulting from the computation error, quantization error, system noise etc., may enter the input channel. In this situation, the steady state value of 0 for the losers and $\pm(\frac{u_{i^*}}{c_1})^{\frac{1}{p+1}}$ for the winner cannot be reached accurately. However, it can be expected that the neural states still converge to the vicinity of the desired values when the additive noise is within certain level. In this part, we explore such a property of the presented model by simulation and compare the robustness of the presented model under different choices of the parameter p (Fig. 4.5).

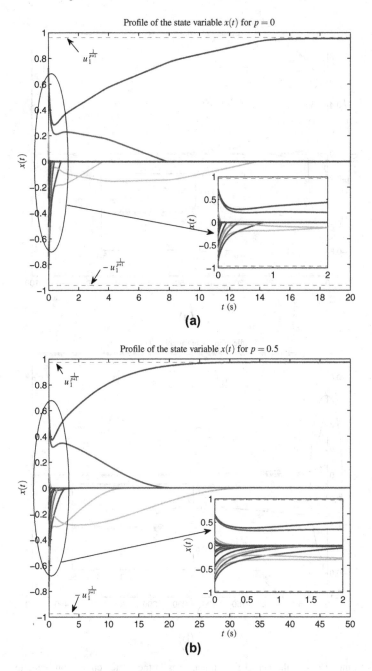

Fig. 4.2 Comparisons of the neural state trajectories in the static competition scenario with 15 neurons under $p = 0$, $p = 0.5$

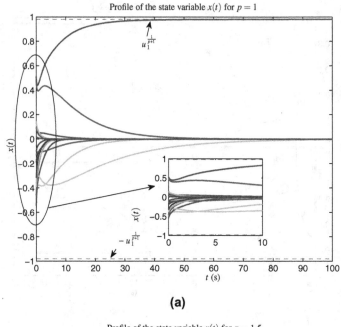

(a)

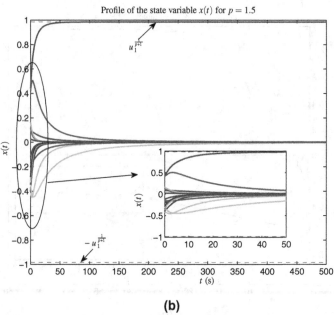

(b)

Fig. 4.3 Comparisons of the neural state trajectories in the static competition scenario with 15 neurons under $p = 1$, $p = 1.5$

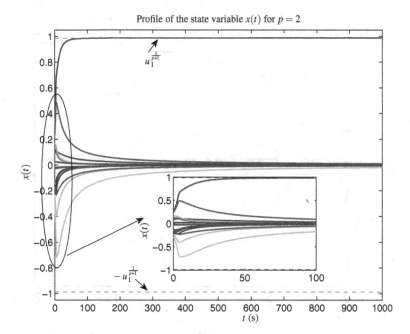

Fig. 4.4 Comparisons of the neural state trajectories in the static competition scenario with 15 neurons under $p = 2$

The noise polluted neural network model considered in this part writes as follows for the ith neuron (Figs. 4.6 and 4.7),

$$\dot{x}_i = c_0(u_i - c_1\|x\|_{p+1}^{p+1})|x_i|^p\mathrm{sgn}(x_i) + v_i, \qquad (4.29)$$

where v_i is a Guassian white noise with zero mean σ^2 variance and is independent with v_j for $i \neq j$. In the simulation, we choose three different noise levels, i.e., $\sigma = 0.1$, $\sigma = 0.5$ and $\sigma = 1$ to evaluate the performance of the presented model. Figure 4.5 through Fig. 4.9 plot the evolution of the neural state trajectories in presence of additive noise under the norm parameter $p = 0$, $p = 1$, $p = 1.5$ and noise level $\sigma = 0.1, \sigma = 0.5, \sigma = 1$. From these figures, we can observe that the neural states are still able to converge to the vicinity of the desired value in the presence of additive noise with a small value of σ, which reveal the robustness of the presented model to noisy inputs. With the increase of σ, the neural state value becomes more and more noisy. For the same level of noise (i.e., the same σ), it can be observed from Fig. 4.5 through Fig. 4.9 that the presented model with a smaller p is less sensitive to the influence of the additive noise. Particularly for $\sigma = 1$ as shown in Fig. 4.5 through Fig. 4.9, the winner can still demonstrate a clear difference from the losers in state value for the cases with $p = 0$ and $p = 1$ while the state values almost mix together for the case with $p = 1.5$ (Fig. 4.8).

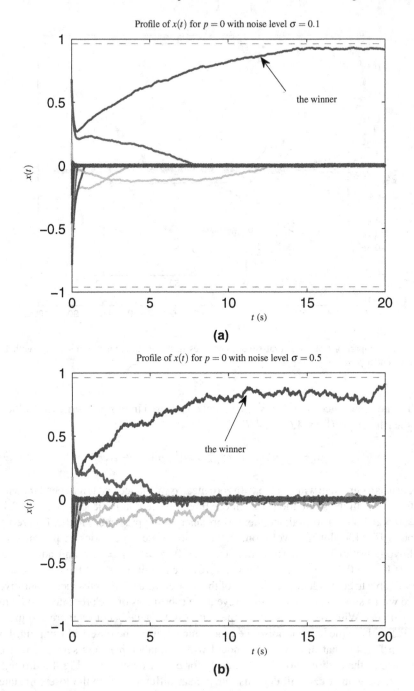

Fig. 4.5 Comparisons of the neural state trajectories of the static competition scenario in presence of additive noise under the norm parameter $p = 0$ and noise level $\sigma = 0.1, \sigma = 0.5$

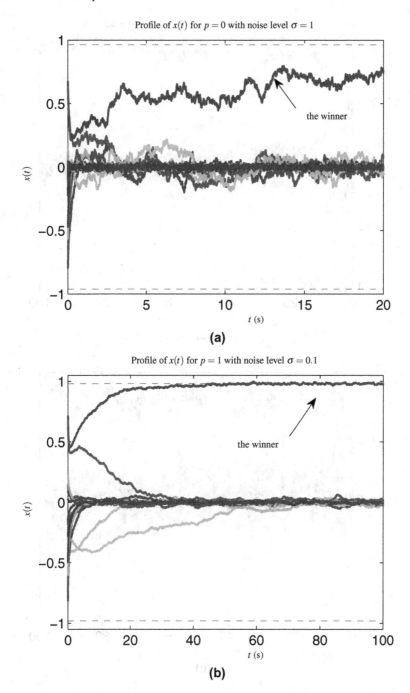

Fig. 4.6 Comparisons of the neural state trajectories of the static competition scenario in presence of additive noise under the norm parameter $p = 0$, $p = 1$ and noise level $\sigma = 0.1$, $\sigma = 1$

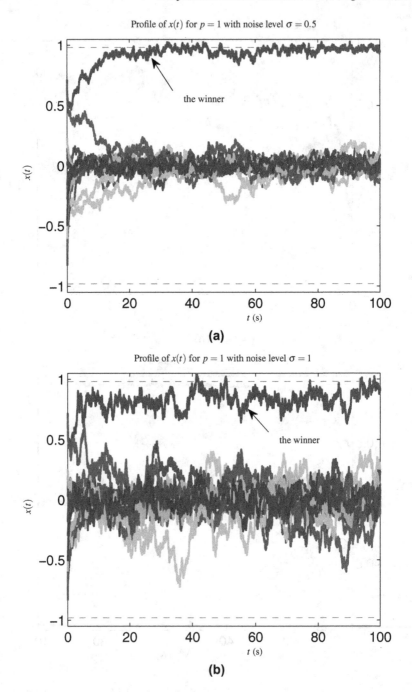

Fig. 4.7 Comparisons of the neural state trajectories of the static competition scenario in presence of additive noise under the norm parameter $p = 1$ and noise level $\sigma = 0.5, \sigma = 1$

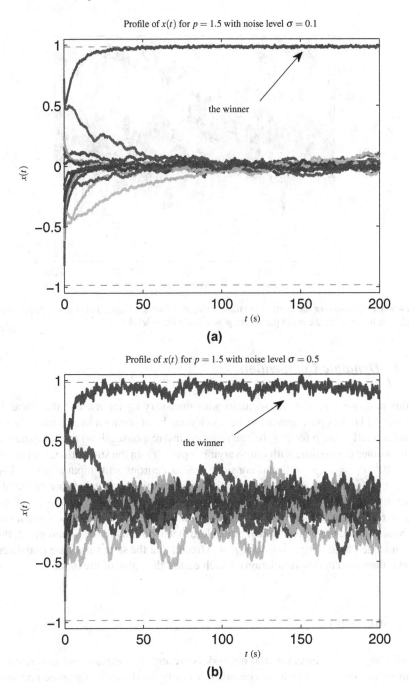

Fig. 4.8 Comparisons of the neural state trajectories of the static competition scenario in presence of additive noise under the norm parameter $p = 1.5$ and noise level $\sigma = 0.1$, $\sigma = 0.5$

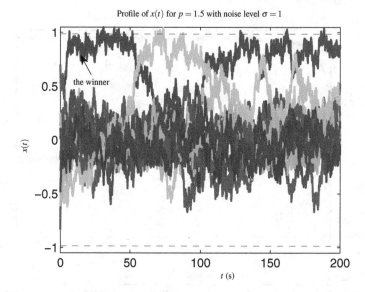

Fig. 4.9 Comparisons of the neural state trajectories of the static competition scenario in presence of additive noise under the norm parameter $p = 1.5$ and noise level $\sigma = 1$

4.6.2   Dynamic Competition

In this part, we consider the scenario with time-varying inputs. For the dynamic system (4.11), the convergence can be accelerated by choosing a large scaling factor c_0 and a small value p for $p \geq 0$. The resulting fast response allows the computation of the winner in real time with time-varying input $u(t)$. In the simulation, we choose $c_0 = 10^4$, $c_1 = 1$, $p = 0$ and consider $n = 3$ neurons with input $u_i(t) = 1 + \sin(2\pi t + \frac{2\pi i}{3})$ for $i = 1, 2, 3$, respectively. The initial state values are randomly generated between 0 and 1. The four input signals and the absolute value of the state variables are plotted in Fig. 4.10. From this figure, we can see the system can successfully find the winner in real time. Note that according to the Theorem 4.1, the output value of the winner is u_{i^*} for $p = 0$ (recall that the state values are initialized greater than zero in this simulation), which equals the value of the input.

4.7   Summary

In this chapter, a recurrent neural network is presented to explain and generate the winner-take-all competition. In contrast to existing models, this dynamic equation features a simple expression and extends the case with Euclidean norm term for neural interaction to the more general p-norm cases. The fact that the state value of the winner converges to be active while the others deactivated is proven theoretically. The

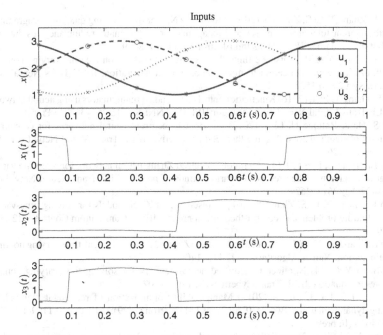

Fig. 4.10 Simulation results for the dynamic winner-take-all competition

convergence rate is discussed based on a local approximation. Simulations with both static inputs and dynamic inputs are performed. Convergence speed and robustness of the presented model against additive noise are also explored by simulation. The results validate the effectiveness of the presented model.

References

1. Dun EA, Ferguson JD, Beveridge CA (2006) Apical dominance and shoot branching. Divergent opinions or divergent mechanisms? Plant Physiol 142(3):812–819
2. Frank R, Cook P (2010) The winner-take-all society: Why the few at the top get so much more than the rest of us. Penguin Books, London
3. Lee D, Itti L, Koch C, Braun J (1999) Attention activates winner-take-all competition among visual filters. Nat Neurosci 2(4):375–381
4. Kurt S, Deutscher A, Crook J, Ohl F, Budinger O, Moeller C, Scheich H, Schulze H (2008) Auditory cortical contrast enhancing by global winner-take-all inhibitory interactions. PLoS ONE 3(3):1–12
5. Enquist M, Ghirlanda S (2005) Neural networks and animal behavior. Princeton University Press, Princeton
6. Jin L, Li S (2017) Distributed task allocation of multiple robots: A control perspective. IEEE Trans Syst Man Cybern Syst pp(99):1–9
7. Li S, Liu B, Li Y (2013) Selective positive-negative feedback produces the winner-take-all competition in recurrent neural networks. IEEE Trans Neural Netw Learn Syst 24(2):301–309

8. Jin L, Zhang Y, Li S, Zhang Y (2016) Modified ZNN for time-varying quadratic programming with inherent tolerance to noises and its application to kinematic redundancy resolution of robot manipulators. IEEE Trans Ind Electron 63(11):6978–6988

9. Jin L, Zhang Y (2015) Discrete-time Zhang neural network for online time-varying nonlinear optimization with application to manipulator motion generation. IEEE Trans Neural Netw Learn Syst 27(6):1525–1531

10. Li S, Zhang Y, Jin L (2016) Kinematic control of redundant manipulators using neural networks. IEEE Trans Neural Netw Learn Syst. doi:10.1109/TNNLS.2016.2574363 (In Press)

11. Li S, He J, Rafique U, Li Y (2017) Distributed recurrent neural networks for cooperative control of manipulators: A game-theoretic perspective. IEEE Trans Neural Netw Learn Syst 28(2):415–426

12. Jin L, Zhang Y, Li S (2016) Integration-enhanced Zhang neural network for real-time varying matrix inversion in the presence of various kinds of noises. IEEE Trans Neural Netw Learn Syst 27(12):2615–2627

13. Jin L, Zhang Y, Li S, Zhang Y (2017) Noise-tolerant ZNN models for solving time-varying zero-finding problems: A control-theoretic approach. IEEE Trans Autom Control 62(2):577–589

14. Jin L, Zhang Y (2016) Continuous and discrete Zhang dynamics for real-time varying nonlinear optimization. Numer Algorithm 73(1):115–140

15. Li S, Li Y (2014) Nonlinearly activated neural network for solving time-varying complex sylvester equation. IEEE Trans Cybern 44(8):1397–1407

16. Jin L, Li S, La H, Luo X (2017) Manipulability optimization of redundant manipulators using dynamic neural networks. IEEE Trans Ind Electron pp(99):1–10. doi:10.1109/TIE.2017.2674624 (In press)

17. Zhang Y, Li S (2017) Predictive suboptimal consensus of multiagent systems with nonlinear dynamics. IEEE Trans Syst Man Cybern Syst pp(99):1–11. doi:10.1109/TSMC.2017.2668440 (In press)

18. Jin L, Zhang Y, Qiu B (2016) Neural network-based discrete-time Z-type model of high accuracy in noisy environments for solving dynamic system of linear equations. Neural Comput Appl. doi:10.1007/s00521-016-2640-x (In press)

19. Li S, Zhou M, Luo X, You Z (2017) Distributed winner-take-all in dynamic networks. IEEE Trans Autom Control 62(2):577–589

20. Jin L, Zhang Y (2015) G2-type SRMPC scheme for synchronous manipulation of two redundant robot arms. IEEE Trans Cybern 45(2):153–164

21. Li S, Cui H, Li Y (2013) Decentralized control of collaborative redundant manipulators with partial command coverage via locally connected recurrent neural networks. Neural Comput Appl 23(1):1051–1060

22. Li S, You Z, Guo H, Luo X, Zhao Z (2016) Inverse-free extreme learning machine with optimal information updating. IEEE Trans Cybern 46(5):1229–1241

23. Khan M, Li S, Wang Q, Shao Z (2016) CPS oriented control design for networked surveillance robots with multiple physical constraints. IEEE Trans Comput-Aided Des Integr Circuits Syst 35(5):778–791

24. Khan M, Li S, Wang Q, Shao Z (2016) Formation control and tracking for co-operative robots with non-holonomic constraints. J Intell Robot Syst 82(1):163–174

25. Benkert C, Anderson DZ (1991) Controlled competitive dynamics in a photorefractive ring oscillator: Winner-takes-all and the voting-paradox dynamics. Phys Rev A 44(1):4633–4638

26. Wang W, Slotine J (2006) Fast computation with neural oscillators. Neurocomputing 69(16–18):2320–2326

27. Rutishauser U, Douglas R, Slotine J (2011) Collective stability of networks of winner-take-all circuits. Neural Comput. 23(3):735–773

28. Xu Z, Jin H, Leung K, Leung Y, Wong C (2002) An automata network for performing combinatorial optimization. Neurocomputing 47(1–4):59–83

29. Liu S, Wang J (2006) A simplified dual neural network for quadratic programming with its k-wta application. IEEE Trans Neural Netw 17(6):1500–1510

30. Hu X, Wang J (2008) An improved dual neural network for solving a class of quadratic programming problems and its k-winners-take-all application. IEEE Trans Neural Netw Learn Syst 19(12):2022–2031

31. Liu Q, Wang J (2008) A one-layer recurrent neural network with a discontinuous activation function for linear programming. Neural Comput. 20(5):1366–1383

32. Khalil H (2002) Nonlinear systems. Prentice-Hall, Englewood Cliffs, NJ

33. Isidori A (1995) Nonlinear control systems, 3rd edn. Springer, New York

34. Mandic D, Chambers J (2000) A normalised real time recurrent learning algorithm. Signal Process 80(9):1909–1916

35. Mandic D, Papoulis E, Boukis C (2003) A normalized mixed-norm adaptive filtering algorithm robust under impulsive noise interference. Proc IEEE Int Conf Acoust 6:333–336

36. Liu Q, Wang J (2011) Finite-time convergent recurrent neural network with a hard-limiting activation function for constrained optimization with piecewise-linear objective functions. IEEE Trans Neural Netw 22(4):601–613

Chapter 5
Distributed Competition in Dynamic Networks

Abstract Consensus has been widely explored in the past years and successfully applied to the design of cooperative control laws and distributed computation paradigms. However, in the light of the great success of consensus in control, the counterpart of consensus, which, instead of mitigating the disagreement, increases the contrasts between dynamic agents in a distributed network, is still missing. The seminal work by Maass [1] (Maass, Neural Comput 12(11), 2519–2535, 2000) proves that weighted averaging, together with the operation of winner-take-all (WTA) organized in a two-layered structure is able to approximate any nonlinear mapping in any desired accuracy. When it comes to distributed networks, Maass's theorem poses great appeal for distributed WTA algorithms provided that the distributed weighted averaging could be addressed using consensus. Unfortunately, as presented in Chaps. 1, 2, 3 and 4, there is no existing distributed WTA algorithm available, which significantly blocks the exhibition of the computational power of WTA over dynamic networks. In this chapter, we make progress along this direction and present the first distributed WTA protocol with guaranteed global convergence. The convergence to the WTA solution is proved rigorously using Lyapunov theory. The theoretical conclusions are supported by numerical validation.

Keywords Winner-take-all competition · Distributed network · Distributed WTA protocol · Global stability · Numerical validation

5.1 Introduction

In past years, dynamic consensus has attracted intensive research attentions, and led to successful solutions of a large variety of distributed computation problems, including distributed coordination of multiple mobile robots [2], decentralized control of redundant manipulators [3]. Despite its great success, consensus algorithm, which updates the state by dynamically mitigating differences among agents, is mostly limited to the modeling of dynamic cooperation. It essentially lacks a mechanism to model dynamic competition in a distributed network, which desires the increase of peer differences and the enhancement of contrasts.

© The Author(s) 2018
S. Li and L. Jin, *Competition-Based Neural Networks with Robotic Applications*,
SpringerBriefs in Applied Sciences and Technology,
DOI 10.1007/978-981-10-4947-7_5

Research in many fields confirms the same importance of competition as that of cooperation in the emergence of complex behaviors. For example, it is revealed in [4] that competition and cooperation plays significant roles in the decision making in market economy. It is demonstrated in [5] that the strategy chosen by the rational politicians consists of cooperation over competition in dealing with international relationships. Recent research in neuroscience found finds that control actions depends on transitory change in patterns of cooperation and competition between brain systems during cognitive control [6]. Due to the fundamental significance of competition in the interaction of multi-agent systems, various models have been presented to capture this competitive nature. Among them, the winner-take-all (WTA) model, which refers to the competition of a group of agents that the one with the largest input finally remains activated while all the other agents are deactivated, has been widely investigated and usually employed to model competition behaviors. Maass proves in [1] that a two-layered network composed of weighted averaging in the first layer and WTA in the second layer is able to approximate any nonlinear mapping in any desired accuracy. Following this results, the dynamic consensus with the capability for the computation of weighted averaging in a distributed way, and a distributed algorithm for the computation of WTA, will be able to constitute any nonlinear mapping in a distributed network. However, to the best of our knowledge, there is no existing algorithm with proved convergence that addresses WTA in a distributed manner. The non-replacement role of distributed WTA in modeling competition behaviors and its potential in complementing consensus to build distributed universal approximators motivate us to design such a dynamic protocol with guaranteed convergence.

Apart from the natures of distributed-storage and high-speed parallel-processing, neural networks can be readily implemented by hardware and thus have been widely applied in various fields, including the description of the WTA competition [7–24]. In previous Chap. 3, it is formulated as a quadratic programming under both linear inequality and equality constraints and solved efficiently in the dual space. Although various models have been presented in Chaps. 1, 2, 3 and 4 as well as references therein to generate WTA, existing work does not take topological constraint into consideration and cannot be directly apply to any connected topology.

In this chapter, we explicitly take the interaction topology of agents into account and present the first fully distributed WTA model to address dynamic competition problems in a network. It extends our previous work in Chaps. 1, 2, 3 and 4 on centralized WTA with a star topology to the general situation with any possible connected topology. To capture the topological property of a graph, algebraic properties on the spectrum of a graph are incorporated into the Lyapunov analysis to prove the global convergence. Different from distributed consensus problem, where a simple Lyapunov function in quadratic form exists, a deliberately designed nonlinear Lyapunov function is employed to prove the asymptotical convergence to WTA.

Notations and Symbols: $\mathbb{R}$ and $\mathbb{R}^+$ denote the field of real numbers and the field of non-negative real numbers, respectively, $\text{diag}(x)$ denotes a diagonal matrix with its diagonal elements from vector x and non-diagonal elements zero, A^T represents the transpose of a matrix A, I represents an identity matrix, $\mathbf{1}$ denotes the vector with all entries equal to 1, e_i denotes a vector with all entries zero except the ith one equal to 1, $\|x\|$ denotes the Euclidean norm of a vector x, the point-wise multiplication of two vectors $x = [x_1, x_2, \ldots, x_n]^T$ and $y = [y_1, y_2, \ldots, y_n]^T \in \mathbb{R}^n$ is defined as $x \circ y = y \circ x = [x_i y_i]$, and the point-wise m-th power of x is defined as $x^m = [x_i^m]$.

5.2 Problem Definition: Distributed WTA on Graphs

Let $G(V, E, W)$ be a weighted undirected graph with n vertices, where $V = \{V_1, \ldots, V_n\}$, $E \subseteq V \times V$, and $W = [W_{ij}] \in \mathbb{R}^{n \times n}$ with $W_{ij} = W_{ji} \geq 0$ for all possible i and j represent the set of nodes, the set of edges and the weighted adjacency matrix, respectively. The element W_{ij}, which lies on the ith row and the jth column in the matrix W, denotes the weight of the edge $E_{ij} \in E$ connecting node V_i and V_j. For $E_{ij} \notin E$, i.e., the case without an edge between node V_i and V_j, $W_{ij} = 0$. The neighbor set of node V_i is defined as $N(i) = \{j \in V, E_{ij} \in E\}$.

Let u_i be the input to node V_i for all i. The node with the largest input among all nodes in the graph, i.e., node $V_{i=\text{argmax}_k\{u_k\}}$, is called the winner on the graph. Across this chapter, we consider the non-trivial case with $\max_k\{u_k\} > \max_{i \neq \text{argmax}_k\{u_k\}}\{u_i\}$, which amounts to exactly a single winner. Let $x_i \in \mathbb{R}$ be a state value associated with node V_i. We say the nodes of the graph G have reached a *WTA* if and only if $x_i = 0$ for $i = \text{argmax}_k\{u_k\}$ (the winner), and $x_i = 0$ for $i \neq \text{argmax}_k\{u_k\}$ (losers). In words, WTA is a status where the winner is active while the losers are deactivated.

In this chapter, we consider using distributed algorithms, particularly dynamic evolutions, to solve WTA on a connected graph. Suppose each node is a dynamic agent with local information exchanging described by the following dynamics,

$$\dot{x}_i = f(x_i, x_{j \in \mathcal{N}(i)}, u_i) \quad i \in V, \tag{5.1}$$

where $x_{j \in \mathcal{N}(i)}$ represents the state variables of those nodes in the neighbor set of node i. Notice that for the class of protocol (5.1), the dynamic evolution of node i's state variable x_i only depends on the state variables of its neighbors $x_{j \in \mathcal{N}(i)}$, the state of itself x_i, and its input u_i, all of which are accessible locally. A prominent advantage of protocol (5.1) is that the information exchanging is limited to direct neighbors, which significantly reduces communication burdens in real applications. A graph $G(V, E, W)$ together with the dynamics (5.1) running on it defines a distributed dynamic network. Now, we are ready to define the distributed WTA problem as follows,

Definition 5.1 Protocol (5.1) asymptotically solves the distributed WTA problem on a graph $G(V, E, W)$ if and only if there exists an asymptotically stable equilibrium $x^* = [x_1^*, x_2^*, \ldots, x_n^*]^T$, satisfying $x_i^* = \text{constant} \neq 0$ for $i = \text{argmax}_k\{u_k\}$ and $x_i^* = 0$ for $i \neq \text{argmax}_k\{u_k\}$.

The distributed WTA problem abstracts dynamics of the global competition via local interactions. This is an essentially challenging problem due to the contrast between the objective to reach *global* maximum over the network and the accessible information limited to the *local* neighborhood of a node. It bears similarity to another challenging problem: the search of the global maximum of a multi-peak function using local information (see Sect. 5.5 for a detailed explanation). In comparison with dynamic consensus protocol [25], which has been widely studied over the past years as a powerful tool for distributed computation, WTA goes to the other extreme than consensus by increasing the disagreement. For consensus problem, including average consensus [25], weighted average consensus [26] or even maximum consensus [27], the disagreement between different nodes decreases to zero with time. In contrast, for the WTA problem, the winner node remains active ultimately while the loser nodes are depressed to zero, which amounts to the increase of the disagreement between the winner and the losers. This unique property poses inherent connections with natural phenomena, such as the contrast gain among overlapping neurons in visual systems [28], the apical dominance among branches in the development of stems [29]. It makes it suitable to model and design distributed competition of multi-agent systems. Although there have been various WTA models, all of them rely information from other nodes than its direct neighbors for the state update of each agent. To the best of our knowledge, the presented protocol in this chapter is the first *fully distributed* dynamic WTA algorithm.

For the derivation of the main results in this chapter, we present some useful preliminaries on graph theory in the following part of this section.

Preliminaries on Graph Theory: The degree of node V_i in a graph $G(V, E, W)$ is defined as $\deg(V_i) = \sum_{j \in \mathcal{N}(i)} W_{ij}$. The degree matrix $\Delta = [\Delta_{ij}]$ is a diagonal matrix with $\Delta_{ij} = 0$ for all $i \neq j$ and $\Delta_{ii} = \deg(V_i)$ for all i. The Laplacian matrix L of a graph G is defined as $L = \Delta - W$. A path between node V_i and node V_j for $i \neq j$ in a graph is a sequence of consecutive edges which connect V_i and V_j via a sequence of neighboring nodes. A graph G is called connected if there always exists a path between any pair of nodes. The Laplacian matrix L of a connected graph is positive semi-definite. Its eigenvalues are all real and satisfy $\lambda_n \geq \lambda_{n-1} \geq \cdots \geq \lambda_2 > \lambda_1 = 0$. The n-dimensional vector $\mathbf{1}$, which has all entries equal to 1, is an eigenvector of L associated with its zero eigenvalue for connected graphs, i.e., $\mathbf{1}^T L = 0$ and $L\mathbf{1} = 0$.

5.3 Distributed WTA Protocol

In this chapter, we present the following dynamic protocol to address the WTA problem on a connected graph $G(V, E, W)$,

$$\dot{x}_i = u_i x_i - 2c_1 x_i y_i - 2c_1^2 x_i^3, \tag{5.2a}$$

$$\dot{y}_i = \sum_{j \in \mathcal{N}(i)} w_{ij}(y_j - y_i + c_1 x_j^2 - c_1 x_i^2), \tag{5.2b}$$

where $x_i \in \mathbb{R}$, $u_i \in \mathbb{R}^+$ and $y_i \in \mathbb{R}$ are the state variable of node i with a random initial value, the non-negative input to node i, and a co-state of it initialized with

$y_i(0) = 0$ for all possible i, respectively, $c_1 \in \mathbb{R}^+$ is a constant, $\mathcal{N}(i)$ denotes the neighbor set of node i, $w_{ij} = w_{ji}$ is the weight of the edge E_{ij} and $w_{ij} > 0$ for $j \in \mathcal{N}(i)$ and $w_{ij} = 0$ for $j \notin \mathcal{N}(i)$. For the consideration of the convergence of protocol (5.2), we make the assumption that $u_i \geq 0$ for all i. In real practice with negative inputs, it is possible to convert the input range into non-negative values without changing the input order of different nodes. One possible way is to first pass the input through a monotone function with the lower bound greater than zero, e.g., the sigmoid function $u_i = \frac{1}{1+e^{-r_i}}$, and then proceed with protocol (5.2) for further processing to screen out the winner. Without the loss of generality, we assume $u_i \geq 0$ for all i across this chapter.

Notice that protocol (5.2) follows the expression of (5.1) by only requiring information exchanging between direct neighbors and thus falls into the distributed protocol class. The dynamics of the network G following protocol (5.2) are written in a compact form as,

$$\dot{x} = u \circ x - 2c_1 x \circ y - 2c_1^2 x^3, \tag{5.3a}$$
$$\dot{y} = -Ly - c_1 L x^2, \tag{5.3b}$$

where $x = [x_1, x_2, \ldots, x_n]^T$, $u = [u_1, u_2, \ldots, u_n]^T$, and $y = [y_1, y_2, \ldots, y_n]^T$ are the state vector, the input vector and the co-state vector, respectively, 'o' is the point-wise multiplication operator, with $x \circ y = [x_i y_i]$, x^m denotes the point-wise mth order power, with $x^3 = x \circ x \circ x = [x_i^3]$ for $m = 3$ and $x^2 = x \circ x = [x_i^2]$ for $m = 2$, $L = [L_{ij}] \in \mathbb{R}^{n \times n}$ is the Laplacian matrix of the graph G, and writes as follows in term of the edge weight w_{ij},

$$L_{ij} = \begin{cases} -w_{ij} & i \neq j \\ \sum_{j \in \mathcal{N}(i)} w_{ij} & i = j. \end{cases} \tag{5.4}$$

Note the definition of $x \circ y$ and x^m for x and y being vectors implies that they are not exchangeable in order with the matrix multiplication, and thus $A(x \circ y) \neq (Ax) \circ y$ in general for a matrix A of an appropriate size. For this consideration, we set the operation of x^m is treated with a highest operational priority when it is composited with other operations, and thus $Lx^2 = L(x^2)$ in (5.3).

5.3.1 Basic Properties

About the protocol (5.2), we observe the following basic properties, which play important role in the derivation of the convergence results.

Lemma 5.1 *For the protocol (5.2) running on a connected graph, the quantity $\sum_{i=1}^{n} y_i$ is invariant with time and it equals zero if y_i is initialized with 0, i.e., $y_i(0) = 0$ for all $i = 1, 2, \ldots, n$.*

This is a direct result exploiting the algebraic property of the Laplacian matrix. The quantity $\sum_{i=1}^{n} y_i = \mathbf{1}^T y$ in vector form. Left multiplying $\mathbf{1}^T$ on both sides of (5.3b) yields $\mathbf{1}^T \dot{y} = \frac{d(\mathbf{1}^T y)}{dt} = 0$ recalling that $\mathbf{1}^T L = 0$, which implies that $\mathbf{1}^T y$ is invariant with time and therefore $\mathbf{1}^T y = \mathbf{1}^T y(0) = 0$ for the case with zero initialization of y.

We also have the following result on the structural symmetry of the presented protocol.

Lemma 5.2 *For protocol (5.2), if (x^*, y^*) for $x^*, y^* \in \mathbb{R}^n$ is an (un)stable equilibrium, then $(-x^*, y^*)$ is an (un)stable equilibrium.*

This result comes from the fact that the system dynamics (5.3) does not change after replacing x with $-x$ in it. In details, conduct the transformation $x' = -x$, $y' = y$ for system (5.3) and it results in the following new dynamics,

$$\dot{x}' = u \circ x' - 2c_1 x' \circ y' - 2c_1^2 x'^3, \tag{5.5a}$$
$$\dot{y}' = -Ly' - c_1 L x'^2. \tag{5.5b}$$

For an (un)stable equilibrium point $(x, y) = (x^*, y^*)$ in the original metric, it is still (un)stable in the new coordinate, i.e., $(x', y') = (-x^*, y^*)$ is (un)stable for (5.5). It is noteworthy that the new system (5.5) is identical to the original one (5.3), meaning that the point $(x', y') = (-x^*, y^*)$ is (un)stable for (5.5) implies $(x, y) = (-x^*, y^*)$ is (un)stable for (5.3). This concludes the claim in Lemma 5.2.

5.4 Convergence Analysis

We conduct convergence analysis in this section to show the global stability of the WTA solution. To this goal, we will first define a nonlinear Lyapunov function and use Lyapunov's direct method to show the global convergence of the system to its equilibrium point set S. To eliminate the other non-WTA solutions, we will further conduct local stability analysis to show that all the other points in set S than the WTA solution are locally unstable. The two parts will constitute the main theory for the global convergence of protocol (5.2) to the WTA solution.

5.4.1 Global Convergence to the Equilibrium Point Set

About the convergence of the presented protocol (5.2), we have the following lemma.

Lemma 5.3 *The dynamic protocol (5.2) on a connected graph with an initialization $y(0) = 0$ asymptotically converges to the equilibrium point set S defined in (5.6) globally.*

$$S = \{(x^*, y^*) \in \mathbb{R}^n \times \mathbb{R}^n, x^* = 0, y^* = 0;$$
$$x^* = \pm\frac{\sqrt{nu_i}}{\sqrt{2c_1}}e_i, \quad y^* = \frac{u_i(1-ne_i)}{2c_1} \quad for\ i = 1, 2, \ldots, n\}. \tag{5.6}$$

Proof The proof of this Lemma includes three steps: 1. The global convergence to an invariant set S. 2. The resolution of the analytical expression of the set S. 3. The correspondence of set S to the equilibrium points of this nonlinear protocol.

Step 1. Global convergence to an invariant set S.

For analysis convenience, we first define the following two auxiliary variables,

$$z = y + c_1 x^2, \tag{5.7a}$$

$$\eta_1 = u \circ x - 2c_1^2 \frac{\|x\|^2 x}{n}, \tag{5.7b}$$

$$\eta_2 = z - \frac{\|x\|^2 \mathbf{1}}{n}, \tag{5.7c}$$

with which the dynamics of z write,

$$\begin{aligned}
\dot{z} &= \dot{y} + c_1\frac{dx^2}{dt} \\
&= -Ly - c_1 Lx^2 + 2c_1 x \circ \dot{x} \\
&= -Lz + c_1 Lx^2 - c_1 Lx^2 + 2c_1 x \circ \dot{x} \\
&= -Lz + 2c_1 x \circ \dot{x}, \tag{5.8}
\end{aligned}$$

where the relation $\frac{dx^2}{dt} = x \circ \dot{x}$ is used. Substituting the expressions of η_1 and η_2 in (5.7) to (5.3a), $\dot{x}$ then become

$$\begin{aligned}
\dot{x} &= u \circ x - 2c_1 x \circ (y + c_1 x^2) \\
&= u \circ x - 2c_1 x \circ z - 2c_1^2 \frac{\|x\|^2 x}{n} + 2c_1^2 \frac{\|x\|^2 \mathbf{1}}{n} \circ x \\
&= \left(u \circ x - 2c_1^2 \frac{\|x\|^2 x}{n}\right) - 2c_1\left(z - c_1\frac{\|x\|^2 \mathbf{1}}{n}\right) \circ x \\
&= \eta_1 - 2c_1 \eta_2 \circ x = \eta_1 - 2c_1 \mathrm{diag}(x)\eta_2. \tag{5.9}
\end{aligned}$$

The equalities $\mathbf{1} \circ x = x$ and $\mathrm{diag}(x)\eta_2 = \eta_2 \circ x$ are employed in the derivation of the second line and the last line of (5.9), respectively. According to the definition of the operators 'diag$(\cdot)$' and '$\circ$', $x^T \dot{x} = \mathbf{1}^T \mathrm{diag}(x)\dot{x} = \mathbf{1}^T(\mathrm{diag}(x)\dot{x}) = \mathbf{1}^T(x \circ \dot{x})$ and further $\mathbf{1}\frac{d\|x\|^2}{dt} = 21x^T\dot{x} = 2\mathbf{1}\mathbf{1}^T(x \circ \dot{x})$. Therefore, the dynamics of η_2 are obtained as,

$$\dot{\eta}_2 = \dot{z} - \frac{c_1 \mathbf{1}}{n} \frac{d\|x\|^2}{dt}$$

$$= -Lz + 2c_1 x \circ \dot{x} - \frac{2c_1}{n} \mathbf{1}\mathbf{1}^T (x \circ \dot{x})$$

$$= -L\eta_2 + 2c_1 (I - \frac{\mathbf{1}\mathbf{1}^T}{n})x \circ (\eta_1 - 2c_1 \text{diag}(x)\eta_2)$$

$$= -\left(L + 4c_1^2 \left(I - \frac{\mathbf{1}\mathbf{1}^T}{n}\right) \text{diag}^2(x)\right) \eta_2 + 2c_1 \left(I - \frac{\mathbf{1}\mathbf{1}^T}{n}\right) x \circ \eta_1$$

$$= -\left(L + 4c_1^2 L_0 \text{diag}^2(x)\right)\eta_2 + 2c_1 L_0 \text{diag}(x)\eta_1, \tag{5.10}$$

where $L_0 = I - \frac{\mathbf{1}\mathbf{1}^T}{n}$. According to Lemma 5.1, we have $\mathbf{1}^T \eta_2 = \mathbf{1}^T z - c_1 \|x\|^2 = \mathbf{1}^T y + c_1 \mathbf{1}^T x^2 - c_1 \|x\|^2 = \mathbf{1}^T y = \mathbf{1}^T y(0) = 0$ (note $\mathbf{1}^T x^2 = \sum_{i=1}^n x_i^2 = \|x\|^2$) and $L_0 \eta_2 = (I - \frac{\mathbf{1}\mathbf{1}^T}{n})\eta_2 = \eta_2$. Thus, Eqs. (5.9) and (5.10) further re-write,

$$\dot{x} = \eta_1 - 2c_1 \text{diag}(x)L_0 \eta_2 \tag{5.11}$$

$$\dot{\eta}_2 = -\left(L + 4c_1^2 L_0 \text{diag}^2(x)L_0 + c_2 \frac{\mathbf{1}\mathbf{1}^T}{n}\right)\eta_2 + 2c_1 L_0 \text{diag}(x)\eta_1, \tag{5.12}$$

where $c_2 > 0$ is a constant.

With the above derivation, we are ready to present the nonlinear Lyapunov function for the convergence analysis. The Lyapunov function V is composed of two parts, i.e., $V = V_1 + V_2$. Define $V_1 = -\frac{1}{2}x^T \text{diag}(u)x + \frac{c_1^2 \|x\|^4}{2n}$. Clearly $V \geq 0$ as $u_i \geq 0$ for all i. Computing the time derivative along the system dynamics yields,

$$\dot{V}_1 = -x^T \text{diag}(u)\dot{x} + \frac{c_1^2}{2n} \frac{d(x^T x)^2}{dt}$$

$$= -(x \circ u)^T \dot{x} + \frac{2c_1^2}{n} \|x\|^2 x^T \dot{x}$$

$$= -\eta_1^T \dot{x} = -\|\eta_1\|^2 + 2c_1 \eta_2^T L_0 \text{diag}(x)\eta_1. \tag{5.13}$$

Additionally, we define $V_2 = \frac{\eta_2^T \eta_2}{2} \geq 0$ and compute its time derivative along the system dynamics as follows,

$$\dot{V}_2 = -\eta_2^T \left(L + 4c_1^2 L_0 \text{diag}^2(x) + c_2 \frac{\mathbf{1}\mathbf{1}^T}{n}\right)\eta_2 + 2c_1 \eta_2^T L_0 \text{diag}(x)\eta_1. \tag{5.14}$$

For the overall Lyapunov function $V = V_1 + V_2 \geq 0$, according to (5.13) and (5.14), it is time derivative is obtained as,

$$\dot{V} = -\|\eta_1\|^2 - 4c_1^2\eta_2^T L_0 \text{diag}^2(x) L_0 \eta_2$$

$$+4c_1\eta_2^T L_0 \text{diag}(x)\eta_1 - \eta_2^T \left(L + c_2\frac{\mathbf{11}^T}{n}\right)\eta_2$$

$$\leq -\|\eta_1\|^2 - 4c_1^2\|\text{diag}(x)L_0\eta_2\|^2$$

$$+4c_1\|\text{diag}(x)L_0\eta_2\| \cdot \|\eta_1\| - \eta_2^T \left(L + c_2\frac{\mathbf{11}^T}{n}\right)\eta_2$$

$$\leq -\left(\|\eta_1\| - 2c_1\|\text{diag}(x)L_0\eta_2\|\right)^2$$

$$-\eta_2^T \left(L + c_2\frac{\mathbf{11}^T}{n}\right)\eta_2. \tag{5.15}$$

According to the spectral theorem [30], the symmetric matrix L can be represented as $L = \lambda_1\frac{\mathbf{11}^T}{n} + \lambda_2\alpha_2\alpha_2^T + \cdots + \lambda_n\alpha_n\alpha_n^T$ where λ_i is the ith eigenvalue of L and α_i is the corresponding eigenvector. Recall that the least eigenvalue of a Laplacian matrix is 0 and the second least eigenvalue is great than 0 for connect graphs, so we have $\lambda_1 = 0$ and $\lambda_i > 0$ for $i \geq 2$. The matrix $L + c_2\frac{\mathbf{11}^T}{n}$ can therefore be represented as $(\lambda_1 + c_2)\frac{\mathbf{11}^T}{n} + \lambda_2\alpha_2\alpha_2^T + \cdots + \lambda_n\alpha_n\alpha_n^T$, from which it is observable that the eigenvalues are $\lambda_1 + c_2 = c_2, \lambda_2, \ldots, \lambda_n$ for $L + c_2\frac{\mathbf{11}^T}{n}$. Accordingly, its least eigenvalue is $\min\{c_2, \lambda_2\} > 0$ and $\eta_2^T \left(L + c_2\frac{\mathbf{11}^T}{n}\right)\eta_2 \geq \min\{\lambda_1 + c_2, \lambda_2\}\eta_2^T\eta_2$. With (5.15), we further get,

$$\dot{V} \leq -\left(\|\eta_1\| - 2c_1\|\text{diag}(x)L_0\eta_2\|\right)^2 - \min\{c_2, \lambda_2\}\|\eta_2\|^2 \leq 0. \tag{5.16}$$

To find the largest invariant set, we set $\dot{V} = 0$ and find its solution from (5.16) as,

$$\eta_1 = 0, \eta_2 = 0. \tag{5.17}$$

According to LaSalle's invariant set principle [31], we conclude that the system dynamics asymptotically evolve to a set S defined by (5.17).

Step 2. Analytical expression of the set S.

We proceed to derive the analytical expressions of the invariant set S. According to (5.7), the equations in (5.17) are equivalent to the following ones,

$$u \circ x - 2c_1^2\frac{\|x\|^2 x}{n} = 0, \tag{5.18}$$

$$z - c_1\frac{\|x\|^2\mathbf{1}}{n} = 0 \rightarrow y + c_1x^2 - c_1\frac{\|x\|^2\mathbf{1}}{n} = 0. \tag{5.19}$$

One simple solution to the above equations is $x = 0$, $y = 0$. For the non-trivial situation $x \neq 0$, note that $\text{diag}(u)x = 2c_1\frac{\|x\|^2 x}{n}$ from (5.18). This is an eigenvalue equation for x, from which it is concluded that $2c_1^2\frac{\|x\|^2}{n}$ is the eigenvalue of $\text{diag}(u)$ associated with the eigenvector x. Since $\text{diag}(u)$ is a diagonal matrix, whose ith eigenvalue is u_i in correspondence to the eigenvector $k_i e_i$ (k_i is a scalar and e_i is a vector with all entry 0 except the ith entry equal 1), the solution of the eigenvalue equation is,

$$x = k_i e_i, \, u_i = \frac{2c_1^2 \|x\|^2}{n}, \text{ for } i = 1, 2, \ldots, n. \tag{5.20}$$

The value of k_i can be obtained by noticing that $\|x\|^2 = k_i^2$ and thus $u_i = \frac{2c_1^2 k_i^2}{n}$. Therefore, $k_i = \pm \frac{\sqrt{nu_i}}{\sqrt{2}c_1}$ and x equals,

$$x = \pm \frac{\sqrt{nu_i}}{\sqrt{2}c_1} e_i. \tag{5.21}$$

Correspondingly, y can be solved from (5.19) as,

$$y = \frac{u_i(1 - ne_i)}{2c_1}. \tag{5.22}$$

In summary, the solution of the invariant set S is obtained as,

$$S = \{(x^*, y^*) \in \mathbb{R}^n \times \mathbb{R}^n, x^* = 0, y^* = 0;$$
$$x^* = \pm \frac{\sqrt{nu_i}}{\sqrt{2}c_1} e_i, \, y^* = \frac{u_i(1 - ne_i)}{2c_1} \text{ for } i = 1, 2, \ldots, n\}. \tag{5.23}$$

Step 3. Correspondence of set S to the equilibrium points.

To show this result, we only need to prove that the equilibrium points are identical to the solution of (5.18) and (5.19). The equilibrium point (x, y) of the presented system (5.3) can be found by solving $\dot{x} = 0$ and $\dot{y} = 0$ in (5.3a) and (5.3b), which yields,

$$0 = u \circ x - 2c_1 x \circ y - 2c_1^2 x^3, \tag{5.24}$$
$$0 = -Ly - c_1 L x^2, \tag{5.25}$$

with the definition of z, the above equations re-write,

$$0 = u \circ x - 2c_1 x \circ z, \tag{5.26}$$
$$0 = -Lz. \tag{5.27}$$

The zero eigenvector of a Laplacian matrix is **1**. Therefore, (5.27) implies $z = k'\mathbf{1}$ where k' is a constant. The fact $z = y + c_1 x^2$ further yields,

$$\mathbf{1}^T(y + c_1 x^2) = k'\mathbf{1}^T\mathbf{1} = k'n. \tag{5.28}$$

According to Lemma 5.1, $1^T y = 0$. Additionally, $1^T x^2 = \sum_{i=1}^{n} x_i^2 = \|x\|^2$. Therefore k' in (5.28) and z are obtained as,

$$k' = \frac{c_1 \|x\|^2}{n}, z = c_1 \frac{\|x\|^2 1}{n}. \tag{5.29}$$

Substituting the expression of z above to (5.26) and (5.27), they are equivalently expressed as,

$$0 = u \circ x - 2c_1^2 x \circ \frac{\|x\|^2 1}{n} = u \circ x - 2c_1^2 \frac{\|x\|^2 x}{n}, \tag{5.30}$$

$$z = c_1 \frac{\|x\|^2 1}{n}, \tag{5.31}$$

which are equivalent to (5.18) and (5.19). This result proves the equivalence of the invariant set S with the equilibrium point set.

About the equilibrium point set S, we have the following remark.

Remark 5.1 Generally speaking, a nonlinear system may demonstrate complex behaviors in its evolution with time. Lemma 5.4 excludes other possibilities for system (5.2) than the ultimate convergence to its equilibrium point set S given in (5.6). It is noteworthy that there are totally $2n + 1$ separate points in the set S, including one special pairs, $x^* = \pm \frac{\sqrt{n u_i}}{\sqrt{2c_1}} e_i$ and $y^* = \frac{u_i(1-ne_i)}{2c_1}$, for $i = \text{argmax}_k \{u_k\}$, which correspond to the WTA solution and appear in pair following Lemma 5.2. Based on Lemma 5.4, we only need to prove the instability of the non-WTA solutions to conclude the global convergence to the WTA solution.

5.4.2 Instability of Non-WTA Solutions

In the previous section, we have derived the analytical expression of the equilibrium points and proved the asymptotical convergence to this set. In this part, we use Lyapunov indirect method to show the instability of the non-WTA solutions and eliminate them from the ultimate stable behavior of the system.

For the convenience of statement, we define the non-WTA solution set S' as a subset of the equilibrium set S to include those non-WTA solution. In expression,

$$S' = \{(x^*, y^*) \in \mathbb{R}^n \times \mathbb{R}^n, x^* = 0, y^* = 0; x^* = \pm \frac{\sqrt{n u_i}}{\sqrt{2c_1}} e_i,$$

$$y^* = \frac{u_i(1 - ne_i)}{2c_1} \text{ for } i = 1, 2, \ldots, n, i \neq \text{argmax}_k \{u_k\}\}$$

$$= S - \{(x^*, y^*) \in \mathbb{R}^n \times \mathbb{R}^n, x^* = \pm \frac{\sqrt{nu_i}}{\sqrt{2c_1}} e_i,$$

$$y^* = \frac{u_i(1 - ne_i)}{2c_1} \text{ for } i = \text{argmax}_k \{u_k\}\}. \tag{5.32}$$

The goal of this subsection is to prove that all points in S' defined in (5.32) are unstable. Towards this goal, we first introduce an auxiliary variable $z = y + c_1 x^2$, and correspondingly a constant $z^* = y^* + c_1 x^{*2}$. With the definition of z, the system dynamics (5.3) can be simplified as

$$\dot{x} = u \circ x - 2c_1 x \circ z, \tag{5.33a}$$
$$\dot{y} = -Ly - c_1 L x^2, \tag{5.33b}$$
$$z = y + c_1 x^2. \tag{5.33c}$$

The dynamics of the errors $\Delta x = x - x^*$, $\Delta y = y - y^*$ and $\Delta z = z - z^*$ are obtained as follows by linearizing (5.33) around the equilibrium points,

$$\dot{\Delta x} = u \circ \Delta x - 2c_1(\Delta z \circ x^* + z^* \circ \Delta x), \tag{5.34a}$$
$$\dot{\Delta y} = -L\Delta y - 2c_1 L(x^* \circ \Delta x), \tag{5.34b}$$
$$\Delta z = \Delta y + 2c_1 x^* \circ \Delta x. \tag{5.34c}$$

There are three classes of equilibrium points in the non-WTA set S' in (5.32). Substituting the expressions of Δz from (5.34c), x^* and y^* from (5.6) into the error dynamics described by (5.34a) and (5.34b), the dynamics of the three classes are respectively obtained as follows:

Class 1 (zero x^*). $x^* = 0$, $y^* = 0$:

$$\dot{\Delta x} = \text{diag}(u)\Delta x, \tag{5.35a}$$
$$\dot{\Delta y} = -L\Delta y. \tag{5.35b}$$

Class 2 (positive x^*). $x^* = \frac{\sqrt{nu_i}}{\sqrt{2c_1}} e_i$, $y^* = \frac{u_i(1-ne_i)}{2c_1}$, for $i \neq \text{argmax}_k\{u_k\}$:

$$\dot{\Delta x} = \text{diag}(u - u_i 1 - 2c_1 nu_i e_i)\Delta x - \sqrt{2c_1 nu_i}\,\text{diag}(e_i)\Delta y, \tag{5.36a}$$
$$\dot{\Delta y} = -\sqrt{2c_1 nu_i}\,L\text{diag}(e_i)\Delta x - L\Delta y. \tag{5.36b}$$

Class 3 (negative x^*). $x^* = -\frac{\sqrt{nu_i}}{\sqrt{2c_1}} e_i$, $y^* = \frac{u_i(1-ne_i)}{2c_1}$, for $i \neq \text{argmax}_k(u_k)$:

$$\dot{\Delta x} = \text{diag}(u - u_i 1 - 2c_1 nu_i e_i)\Delta x + \sqrt{2c_1 nu_i}\,\text{diag}(e_i)\Delta y, \tag{5.37a}$$
$$\dot{\Delta y} = -\sqrt{2c_1 nu_i}\,L\text{diag}(e_i)\Delta x - L\Delta y. \tag{5.37b}$$

Δx and Δy in Class 1 are decoupled in their dynamics and the system matrix for Δx, which is diag(u), is positive definite and thus the equilibrium point in Class 1 is unstable.

For Class 2, it is noteworthy e_i has a single non-zero element on its ith element. As a result, all elements of Δx except the ith one are decoupled in their dynamics. In equation, the dynamics of the jth element of Δx writes,

$$\dot{\Delta x}_j = \begin{cases} (u_j - u_i)\Delta x_j & \text{if } j \neq i \\ -2c_1 n u_j \Delta x_j - \sqrt{2c_1 n u_j} \Delta y_j & \text{if } j = i \end{cases} \tag{5.38}$$

Particularly, we consider the case $j = j = \text{argmax}_k\{u_k\}$ in this situation. Noting $i \neq \text{argmax}_k\{u_k\}$, the dynamics of x_j for $j = \text{argmax}_k\{u_k\} \neq i$ is as follows,

$$\dot{\Delta x}_j = (u_j - u_i)\Delta x_j = (u_{max} - u_i)\Delta x_j. \tag{5.39}$$

Clearly, Δx_j grows to infinity with time since $u_{max} - u_i > 0$ for $i \neq \text{argmax}_k\{u_k\}$. Therefore, the overall system in this case is unstable.

The instability of Class 3 can be directly concluded employing Lemma 5.2 by noticing that solutions of Class 3 can be expressed as $(-x^*, y^*)$ in terms of the solution (x^*, y^*) of Class 2.

So far, we have proved the instability of all points in the non-WTA solution set S' and the theoretical conclusion is summarized in the following lemma.

Lemma 5.4 *For the distributed WTA protocol (5.2) on connected graphs, all points in the non-WTA equilibrium point set S' defined in (5.32) are unstable.*

About Lemma 5.4, we have the following remark.

Remark 5.2 The local instability of points in the non-WTA solution set S', which is a subset of the equilibrium point set S, implies that any small perturbation imposed on the state variables of protocol (5.2) will lead to a permanent deviation of the state values from the original ones.

5.4.3 Global Stability of the WTA Solution

In the past two subsections, we have proved the global convergence to the equilibrium point set S, and the instability of equilibrium points in the non-WTA solution set S'. Based on them, we will draw the conclusion of the global stability of the WTA solution.

As mentioned in Remark 5.1, the global convergent set S is composed of $2n + 1$ separate equilibrium points. Due to the continuous nature of the state variable x and the co-state variable y relative to time t, we conclude that there must exist at least one stable equilibrium point for the presented protocol and the system converge to those stable equilibriums ultimately. Lemma 5.4 proves the instability of non-WTA equilibrium points (set S') and leaves two possible stable equilibrium points, i.e., $S - S' = \{(x^*, y^*) \in \mathbb{R}^n \times \mathbb{R}^n, x^* = \pm \frac{\sqrt{nu_i}}{\sqrt{2}c_1} e_i, y^* = \frac{u_i(1-ne_i)}{2c_1}$ for $i = \text{argmax}_k\{u_k\}\}$. Notice that this set include two points in the form of (x^*, y^*) and $(-x^*, y^*)$. Recalling Lemma 5.2, which proves the same stability property for the point (x^*, y^*) and $(-x^*, y^*)$, it is concluded that there are only two possibilities: the points in $S - S'$ are either both stable or both unstable. Assume the second possibility, i.e., both points in set $S - S'$ are unstable. Note that all points in the set S are unstable in this case due to the fact that all points in S' are unstable. It turns out to a contradiction with Lemma 5.4, which proves the global convergence to S. As as result, the two points in the set $S - S'$ have to be both stable. In summary, we have the following theorem.

Theorem 5.1 *Protocol (5.2) with a random initialization of x and an initialization of $y(0) = 0$ asymptotically solves the distributed WTA problem on connected graphs. x and y globally converge to $x^* = \pm \frac{\sqrt{nu_i}}{\sqrt{2}c_1} e_i$, $y^* = \frac{u_i(1-ne_i)}{2c_1}$ with $i = \text{argmax}_k\{u_k\}$.*

About the criteria to recognize the winner and losers using the state variable x, we have the following remark.

Remark 5.3 Until now, we have proved the stability of the presented distributed WTA protocol. For an agent in the dynamic network, say the ith one, it can identify whether it is a winner or a loser according to the its state value x_i: it is the winner if and only if $\lim_{t \to \infty} x_i \neq 0$ and a loser if and only if $\lim_{t \to \infty} x_i = 0$. In practice, this criteria can be approximated by comparing the value $\|x_i\|$ with a small threshold $\varepsilon > 0$ after running the protocol for an enough long time. In addition, for the winner $i = \text{argmax}_k\{u_k\}$, $\lim_{t \to \infty} x_i = x_i^* = \pm \frac{\sqrt{nu_i}}{\sqrt{2}c_1}$. As a by-product, the number of nodes on the graph can be estimated by the winner using $n = \frac{2c_1^2 x_i^{*2}}{u_i}$.

About the co-state variable y, we have the following remark.

Remark 5.4 According to Theorem 5.1, the co-state y converges to $y^* = \frac{u_i(1-ne_i)}{2c_1}$ with $i = \text{argmax}_k\{u_k\}$. For a loser agent $j \neq i$, its co-state value is $y_j^* = \frac{u_j}{2c_1}$, and conveys the value of the winner input u_i. For the winner agent i, the co-state value is $y_i^* = \frac{u_i(1-n)}{2c_1} < 0$ for $n \geq 2$. For large scale network with $n \gg 1$, the value of the winner co-state y_i^* is much greater than the loser co-state value y_j^* for $j \neq i$ ($\lim_{n \to \infty} \frac{y_i^*}{y_j^*} = -n$), and therefore the co-state value y also demonstrates WTA approximately for enough large n.

5.5 Numerical Validation

In this part, we use numerical simulations to validate the theoretical conclusions drawn in the chapter.

We first consider a simple case with 10 agents to illustrate the dynamic competition behaviors. The communication graph topology is shown in Fig. 5.1. In the simulation, the coefficient c_1 is chosen as $c_1 = 1$ and the weight of each link in the graph is set as 100. The input value v is randomly set as $v = [0.5673, 0.4285, 0.7645, 0.3673, 0.4333, 0.8203, 0.6813, 0.3004, 0.4293, 0.8900]$. The initial value of the state variable x is randomly chosen as $[-3.2873, -4.4980, -9.1094, -8.1221, -1.8001, 6.3378, 7.4103, -9.5489, 4.5435, 6.9602]$. For the co-state variable y, it is initialized as zero. Figure 5.2 shows the dynamic evolution of the state variable x and the co-state variable y with time. Due to the nonlinear interactions, some agents demonstrate fluctuations in their transient to the steady state (e.g., see the yellow curve in Fig. 5.2a, which corresponds to the node 6). Ultimately, the node 10 (the red curve in Fig. 5.2a and b), which have the largest input value $u_{10} = 0.8900$, suppresses activities of all the other agents to zero, and remains active at the end of the simulation. The ultimate value of the winner at the end of the simulation run, i.e., $x_{10}(50)$, equals to 2.1103, which is approximately equal to $\frac{\sqrt{nu_{10}}}{\sqrt{2c_1}} = 2.1095$. This in turn validates the theoretical conclusion on the ultimate value of the winner.

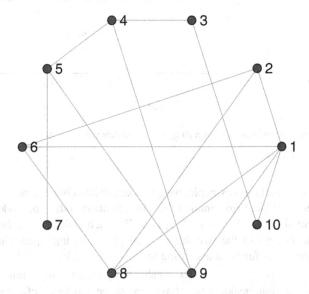

Fig. 5.1 The topology of a communication graph with 10 nodes, where the solid nodes represent the dynamic agents, the edges between nodes represent the information exchanging connection

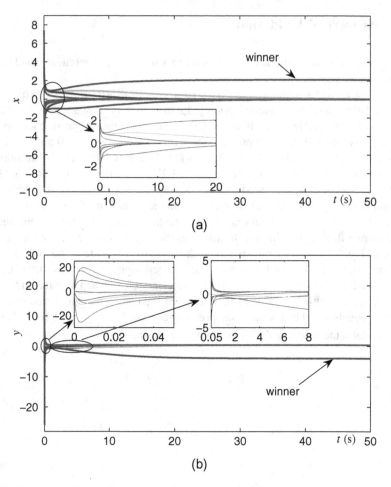

Fig. 5.2 The time profiles in the case with 10 agents: **a** The state variable x; **b** The co-state variable y

In the second simulation example, we consider a relatively large network involving 200 nodes with local maximum values in neighborhoods. The nodes are randomly distributed in a $[-2, 2] \times [-2, 2]$ area. For each node, say the ith one with the coordinate (x_i, y_i) in the two-dimensional plane, its input value u_i is computed using the peaks function according to $u_i = 3(1 - x_i)^2 e^{-x_i^2 - (y_i+1)^2} - 10(\frac{x}{5} - x_i^3 - y_i^5) e^{-x_i^2 - y_i^2} - \frac{1}{3} e^{-(x_i+1)^2 - y_i^2} + 10$. As shown by the contour of this function in Fig. 5.3, it has multiple peaks in the considered range and local information based maximization algorithms, e.g., gradient ascend approaches, may finally drop into a local optima ultimately. For the presented distributed WTA algorithm, each agent only relies information from its neighbor agents, i.e., local information, to update its state. The similarity between the maximization problem and the WTA problem

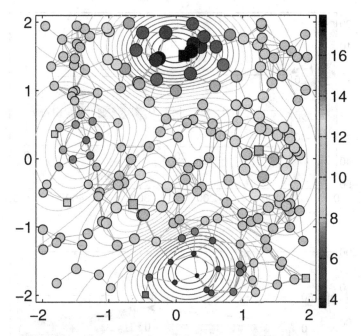

Fig. 5.3 The topology of a communication graph with 200 nodes using inputs generated by peaks function. In this figure, each node represents a dynamic agent, and the edges between nodes represent the information exchanging connection. A *square* shape is used for the representation of those nodes with the largest inputs in its neighborhood. For other nodes, they are represented with *solid circles*. Both the pseudo-color and the size of the nodes indicate the values of inputs to each agent (The larger the node size is, the greater the input is. The *color-bar aside* shows the correspondence between the pseudo-color and the node size.). The level curves in this figure denote the contour of the function employed to generate the input data of each agent

may lead us to the conclusion that the presented distributed WTA, when employed to identify the agent with the maximum input value (the winner), may also suffer from local maxima problem. To test the results with the presented algorithm, we choose the same value of c_1 and the same weight for each edge as in the simulation with 10 agents. With the inputs chosen using the peaks function, there are totally 7 nodes with local maximum input values in their neighborhood (as denoted by the square in Fig. 5.3 for those nodes). Among all the nodes, node 130 has the largest input globally. With a random initialization of x and initialization of y from zero, a typical simulation run is shown in Fig. 5.4a for the time evolution of x, and Fig. 5.4b for the time evolution of y, respectively. Counter-intuitively, as reflected from the simulation results, all agents become in-active in their states except the 130th node, with a state value of -42.2818 remaining active at the end of the simulation for 500 seconds. In other words, different from local information based maximization algorithms, the presented algorithm successfully identifies the agent with the globally maximum input (the winner), only relying on local information from neighbors.

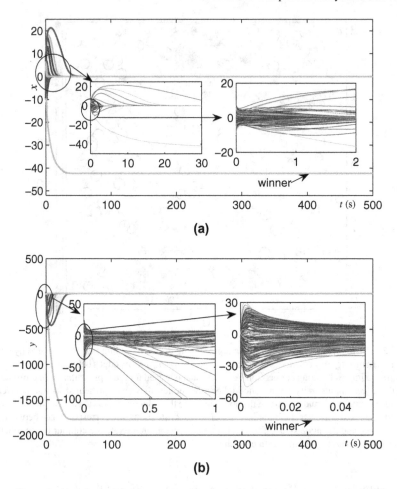

Fig. 5.4 The time profiles in the case with 200 agents using peaks function for the generation of inputs: **a** The state variable x; **b** The co-state variable y

To show the effectiveness of the presented WTA algorithm in the presence of more local maximum input values, we consider another simulation example with the communication graph topology as shown in Fig. 5.5 and random inputs (the input value to each agent is indicated using the size of the node and its pseudo-color). Due to the randomness in the input values, there exists more local maximum values for the inputs to agents in their neighborhoods. In total, there are 29 local maximum inputs. With the same choices of the parameters as in the example with peaks function to generate inputs, and a random initialization of the state variable x and zero initialization of the co-state y, we observer the winner-take-all behavior in the state variable x and the co-state y as shown in Fig. 5.6. From Fig. 5.6a, it is clear that all the state variables are gradually suppressed to zero except one state

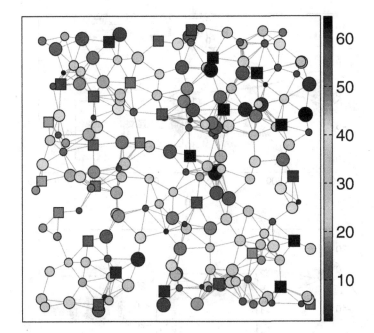

Fig. 5.5 The topology of a communication graph with 200 nodes using random inputs. In this figure, each node represents a dynamic agent, and the edges between nodes represent the information exchanging connection. A *square* shape is used for the representation of those nodes with the largest inputs in its neighborhood. For other nodes, they are represented with *solid circles*. Both the pseudo-color and the size of the nodes indicate the values of inputs to each agent (The larger the node size is, the greater the input is. The *color-bar aside* shows the correspondence between the pseudo-color and the node size)

variable remaining active after a period of transient. The simulation data shows that the active state at the end is associated with the 102th agent, and the final state value is -9.9913, which approximately equals $-\frac{\sqrt{nu_{102}}}{\sqrt{2}c_1} = -\frac{\sqrt{200 \times 0.9983}}{\sqrt{2} \times 1} = -9.9914$. This results validates the theoretical conclusion drawn in this chapter.

5.6 Summary

In this chapter, we defined the distributed WTA problem on connected graphs and presented a dynamic protocol for the solution. As the presented protocol includes multiple equilibriums, we first proved the convergence of the equilibrium point set and then the instability of non-WTA equilibrium points, and finally concluded the asymptotical convergence of the system dynamics to the WTA solution. Simulations were conducted to validate the theoretical conclusions.

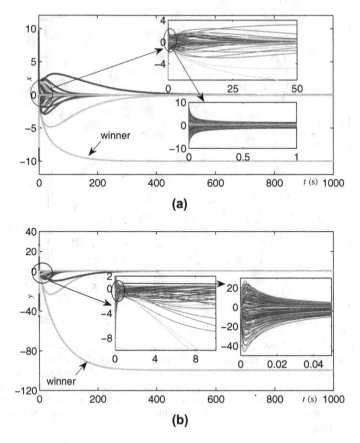

Fig. 5.6 The time profiles in the case with 200 agents using random inputs: **a** The state variable x; **b** The co-state variable y

References

1. Maass W (2000) On the computational power of winner-take-all. Neural Comput 12(11):2519–2535
2. Li S, Kong R, Guo Y (2014) Cooperative distributed source seeking by multiple robots: algorithms and experiments. IEEE/ASME Trans Mech 19(6):1810–1820
3. Li S, Chen S, Liu B, Li Y, Liang Y (2013) Decentralized kinematic control of a class of collaborative redundant manipulators via recurrent neural networks. Neurocomputing 91:1–10
4. Coughlan A (1985) Competition and cooperation in marketing channel choice: theory and application. Mark Sci 4(2):110–129
5. Glaser C (2010) Rational theory of international politics: the logic of competition and cooperation. Princeton University Press, Princeton
6. Cocchi L, Zalesky A, Fornito A, Mattingley J (2013) Dynamic cooperation and competition between brain systems during cognitive control. Trends Cognit Sci 17(10):493–501
7. Li S, Liu B, Li Y (2013) Selective positive-negative feedback produces the winner-take-all competition in recurrent neural networks. IEEE Trans Neural Netw Learn Syst 24(2):301–309

8. Jin L, Zhang Y, Li S, Zhang Y (2016) Modified ZNN for time-varying quadratic programming with inherent tolerance to noises and its application to kinematic redundancy resolution of robot manipulators. IEEE Trans Ind Electr 63(11):6978–6988

9. Jin L, Zhang Y (2015) Discrete-time Zhang neural network for online time-varying nonlinear optimization with application to manipulator motion generation. IEEE Trans Neural Netw Learn Syst 27(6):1525–1531

10. Li S, Zhang Y, Jin L (2016) Kinematic control of redundant manipulators using neural networks. IEEE Trans Neural Netw Learn Syst. doi:10.1109/TNNLS.2016.2574363 (In Press)

11. Li S, He J, Rafique U, Li Y (2017) Distributed recurrent neural networks for cooperative control of manipulators: A game-theoretic perspective. IEEE Trans Neural Netw Learn Syst 28(2):415–426

12. Jin L, Zhang Y, Li S (2016) Integration-enhanced Zhang neural network for real-time varying matrix inversion in the presence of various kinds of noises. IEEE Trans Neural Netw Learn Syst 27(12):2615–2627

13. Jin L, Zhang Y, Li S, Zhang Y (2017) Noise-tolerant ZNN models for solving time-varying zero-finding problems: a control-theoretic approach. IEEE Trans Autom Control 62(2):577–589

14. Jin L, Zhang Y (2016) Continuous and discrete Zhang dynamics for real-time varying nonlinear optimization. Numer Algorithms 73(1):115–140

15. Li S, Li Y (2014) Nonlinearly activated neural network for solving time-varying complex sylvester equation. IEEE Trans Cybern 44(8):1397–1407

16. Jin L, Li S, La H, Luo X (2017) Manipulability optimization of redundant manipulators using dynamic neural networks. IEEE Trans Ind Electr (99):1–10. doi:10.1109/TIE.2017.2674624 (In press)

17. Zhang Y, Li S (2017) Predictive suboptimal consensus of multiagent systems with nonlinear dynamics. IEEE Trans Syst Man Cybern Syst (99):1–11. doi:10.1109/TSMC.2017.2668440 (In press)

18. Jin L, Zhang Y, Qiu B (2016) Neural network-based discrete-time Z-type model of high accuracy in noisy environments for solving dynamic system of linear equations. Neural Comput Appl. doi:10.1007/s00521-016-2640-x (In press)

19. Li S, Zhou M, Luo X, You Z (2017) Distributed winner-take-all in dynamic networks. IEEE Trans Autom Control 62(2):577–589

20. Jin L, Zhang Y (2015) G2-type SRMPC scheme for synchronous manipulation of two redundant robot arms. IEEE Trans Cybern 45(2):153–164

21. Li S, Cui H, Li Y (2013) Decentralized control of collaborative redundant manipulators with partial command coverage via locally connected recurrent neural networks. Neural Comput Appl 23(1):1051–1060

22. Li S, You Z, Guo H, Luo X, Zhao Z (2016) Inverse-free extreme learning machine with optimal information updating. IEEE Trans Cybern 46(5):1229–1241

23. Khan M, Li S, Wang Q, Shao Z (2016) CPS oriented control design for networked surveillance robots with multiple physical constraints. IEEE Trans Comput Aided Des Integr Circuits Syst 35(5):778–791

24. Khan M, Li S, Wang Q, Shao Z (2016) Formation control and tracking for co-operative robots with non-holonomic constraints. J Intell Robot Syst 82(1):163–174

25. Olfati-Saber R, Murray R (2004) Consensus problems in networks of agents with switching topology and time-delays. IEEE Trans Autom Control 49(9):1520–1533

26. Xiao L, Boyd S, Lall S (2005) A scheme for robust distributed sensor fusion based on average consensus. In 4th international symposium on information processing in sensor networks, pp 63–70

27. Iutzeler F, Ciblat P, Jakubowicz J (2012) Analysis of max-consensus algorithms in wireless channels. IEEE Trans Signal Process 60(11):6103–6107

28. Lee D, Itti L, Koch C, Braun J (1999) Attention activates winner-take-all competition among visual filters. Nat Neurosci 2(4):375–381

29. Dun EA, Ferguson JD, Beveridge CA (2006) Apical dominance and shoot branching. Divergent opinions or divergent mechanisms? Plant Physio 142(3):812–819
30. Meyer C (2000) Matrix analysis and applied linear algebra. SIAM, Philadelphia, PA, USA
31. Khalil H (2002) Nonlinear systems. Prentice Hall, USA

Chapter 6
Competition-Based Distributed Coordination Control of Robots

Abstract In this chapter, as a application of the competition-based models investigated in previous chapters, the problem of dynamic task allocation in a distributed network of redundant robot manipulators for path-tracking with limited communications is investigated, where k fittest ones in a group of n redundant robot manipulators with $n > k$ are allocated to execute an object tracking task. The problem is essentially challenging in view of the interplay of manipulator kinematics and the dynamic competition for activation among manipulators. To handle such an intricate problem, a distributed coordination control law is developed for the dynamic task allocation among multiple redundant robot manipulators with limited communications and with the aid of a consensus filter. In addition, a theorem and its proof are presented for guaranteeing the convergence and stability of the proposed distributed control law. Finally, an illustrative example is provided and analyzed to substantiate the efficacy of the proposed control law.

Keywords Winner-take-all competition · Motion generation and control · Redundant robot manipulators · Limited communications · Manipulator kinematics

6.1 Introduction

Cooperation and competition play fundamental roles in the interaction of multi-agent systems by enhancing the harmony and flexibility of the group and make multi-agent systems advantageous in avoiding predators, foraging, energy conservation, and besieging and capturing preys [1, 2]. Therefore, distributed cooperation of multi-agent systems, such as the group of mobile robots, unmanned surface vessels, autonomous underwater vehicles, or unmanned aerial vehicles, has received considerable attention [1, 2]. In such a distributed system, each agent receives the information from its neighboring agents and then responds according to the consensus control protocol. Consensus algorithms, as modeling of cooperation of a multi-agent system, update the state by mitigating differences among agents involved. They have been widely investigated and employed in many distributed problems, such as [1, 2] as well as the references therein.

© The Author(s) 2018

S. Li and L. Jin, *Competition-Based Neural Networks with Robotic Applications*,
SpringerBriefs in Applied Sciences and Technology,
DOI 10.1007/978-981-10-4947-7_6

Recently, robotics as well as other smart agents have been playing more and more significant roles in scientific researches and engineering applications [3–11]. Note that redundant robot manipulators play an important role in various fields [5, 8–11]. One of the fundamental issues in operating redundant robot manipulators is the kinematic redundancy resolution problem. That is, given the desired Cartesian trajectories of the manipulator's end-effector, the corresponding joint trajectories need to be generated in real time. Compared with a single redundant robot, a system composed of multiple redundant manipulators has much more dexterity and flexibility, and can complete more complex tasks. Motivated by these, it makes sense to investigate the coordination of multiple redundant manipulators for executing various task.

As observed in many fields, competition is of the same importance as cooperation in the emergence of complex behaviors [12]. The exploitation of competition among multiple agents provides a possibility to investigate task allocation in a multi-agent system. However, to the best of our knowledge, there is no existing scheme with proved stability that addresses the problem of dynamic task allocation in a system of multiple redundant robot manipulators in a distributed manner. The problem is essentially challenging in view of the contrast between the objective to allocate the task to the k fittest manipulators from a global perspective and the accessible information limited to the local neighborhood of a manipulator as well as the integration of robot kinematics. In comparison with the dynamic consensus problem, which is widely investigated over the last decade as a powerful tool for distributed cooperation of multi-agent systems, it is required to increase the contrast between task executors (the winners) and the rest ones (the losers) for task allocation.

The k-winners-take-all (k-WTA) strategy, which performs the selection of the k competitors whose inputs are larger than the rest ones, has been presented and investigated to describe and capture this competitive nature [13–22]. Apart from the natures of distributed-storage and high-speed parallel-processing, neural networks can be readily implemented by hardware and thus have been widely applied in various fields, including the k-WTA strategy [13–16]. It has been proven in [13] that a two-layered network composed of weighted averaging in the first layer and WTA in the second layer is able to approximate any nonlinear mapping in any desired accuracy. By following optimization based formulation, WTA problem can be modeled as a constrained convex quadratic programming (QP) problem, and then gradient descent or projected gradient descent is employed to get the corresponding dynamic equations for online solution of the problem [16]. It is worth pointing out that many algorithms have been presented for online solution of QP with different emphases, such as models presented in [4, 9, 23] as well as references therein.

Although extensive achievements have been gained for the control of a single redundant manipulator, the research on the redundancy resolution of multiple manipulators is far from up-to-date, which severely restrict its applications in practical and academical research. In [9], a simultaneous repetitive motion planning and control (SRMPC) scheme is designed for synchronous manipulation of two redundant robot arms. However, this paper only considers a dual arm system and the generalization to a network of manipulators remains unclear. Cooperative kinematic control of multiple manipulators is investigated in [10], which uses distributed recurrent

Table 6.1 Comparisons among different schemes for redundancy resolution of manipulators

	Manipulator numbers	Distributed versus centralized	Topology	All connected to command center	Ability to task allocation	Singularity avoidance
This chapter	Multiple	Distributed	Neighbor-to-Neighbor	No	Yes	Yes
Papers [4, 5]	Single	NA	NA	NA	NA	No
Paper [9]	Two	Centralized	NA	Yes	No	No
Paper [10]	Multiple	Distributed	Neighbor-to-Neighbor	No	No	No
Paper [8]	Multiple	Distributed	Star	Yes	No	No
Paper [11]	Multiple	Distributed	Tree*	No	No	No

*Note that NA means that the item does not apply to the algorithm in the associated papers

neural networks and provides a tractable way to extend existing results on individual manipulator control to the scenario with the coordination of multiple manipulators. Comparisons among existing schemes for manipulator redundancy resolution are summarized in Table 6.1. To the best of the authors' knowledge, there is no systematic solution on dynamic task allocation in distributed coordination of multiple robot manipulators.

In this chapter, as shown in Fig. 6.1, a new coordination behavior is first defined in a competition manner for path-tracking via multiple redundant robot manipulators, in which only the fittest ones are allocated the task and with their end-effectors being activated to track the desired trajectory generated by a moving target while the rest ones keep deactivated and unmoved. This is quite different from the existing cooperation control of multiple redundant robot manipulators, which often requires all redundant robot manipulators to execute the task together. With the aid of a consensus filter, a distributed coordination control law for the dynamic task allocation among multiple redundant manipulators is proposed, which can solve such an intricate problem without relying any global information.

6.2 Preliminary and Problem Formulation

In order to lay a basis for further investigation, the preliminary on the redundancy resolution of the redundant robot manipulator, the problem formulation on dynamic task allocation and the existing k-winners-take-all (k-WTA) model are first presented in this section.

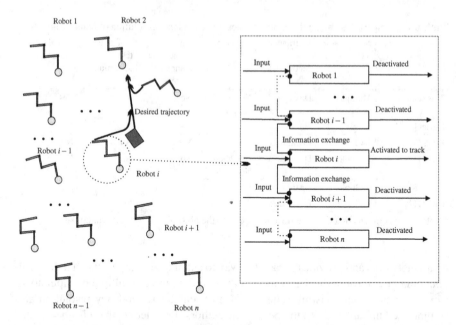

Fig. 6.1 With limited communications among redundant manipulators, how to allocate the path-tracking task dynamically to the k redundant manipulators with their end-effectors being activated to track the desired trajectory generated by the moving target and with the rest $n - k$ ones kept unmoved? With our deliberately designed coordination control law in this chapter, only the k fittest redundant manipulators, in terms of distance from its end-effector to the target, are activated to execute the task while the rest ones keep unmoved

6.2.1   Redundant Robot Manipulator

For the ith redundant robot manipulator with p joints (or p DOF, degrees-of-freedom), the end-effector position vector $r_i \in \mathbb{R}^m$ can be described by the following equation:

$$r_i = f_i(\theta_i), \tag{6.1}$$

where $\theta_i \in \mathbb{R}^p$ refers to angles of the p joints, and $f_i(\cdot)$ is a differentiable nonlinear function with a known structure and parameters for the ith manipulator. In addition, $J_i \in \mathbb{R}^{m \times p}$ is the Jacobian matrix defined as $J_i = \partial f_i(\theta_i)/\partial \theta_i$. Note that r_i is expected to track the desired path r_d, i.e., $r_i = f_i(\theta_i) \rightarrow r_d$. Based on the gradient descent method, an inverse-free control scheme can be designed for the ith redundant robot manipulator to track the desired path via the following steps.

First, define a norm-based energy function:

$$\varepsilon_i = \|r_d - f_i(\theta_i)\|_2^2/2, \tag{6.2}$$

where $\| \cdot \|_2$ denotes the two-norm of a vector.

Second, a rule is designed to evolve along a descent direction of this energy function until the minimum point is reached. The typical descent direction is the negative gradient of ε_i, i.e.,

$$- \partial \varepsilon_i / \partial \theta_i = J_i^{\mathrm{T}} (r_{\mathrm{d}} - f_i(\theta_i)), \tag{6.3}$$

where superscript ^{T} denotes the transpose of a matrix or vector. Then we combine the aforementioned negative gradient (6.3) and the following gradient dynamics design formula :

$$\dot{\theta}_i = -c_0 \partial \varepsilon_i / \partial \theta_i, \tag{6.4}$$

where parameter $c_0 > 0$ denotes the feedback gain used to scale the convergence rate.

Finally, we thus have the following inverse-free control scheme for solving the inverse kinematics problem of the ith redundant robot manipulator:

$$\dot{\theta}_i = c_0 J_i^{\mathrm{T}} (r_{\mathrm{d}} - f_i(\theta_i)). \tag{6.5}$$

Evidently, the above control scheme does not require the Jacobian inversion operation usually existing in pseudoinverse-based solution.

6.2.2 Problem Definitions and Assumptions

We present definitions on the communication graph and communication topology.

Communication graph is the graph with the nodes being redundant manipulators and the edges being communication links. Moreover, $\mathbb{N}(i)$ is used to denote a set of redundant manipulators with communication links to the ith redundant manipulator, which is to represent the neighbor set of the ith redundant manipulator on the communication graph.

Moveover, the definition on communication topology of limited communications is presented as follows [6].

Assumption *Limited communication*, with each redundant manipulator as a node and the communication link between one-hop neighboring robots as edges, is of the communication topology being a connected undirected graph. We use $j \in \mathbb{N}(i)$ denoting the neighbor set of the ith redundant manipulator in the communication graph.

In this chapter, the dynamic task allocation in the coordination of multiple redundant manipulators for path-tracking in a competition manner with limited communication topology is considered, which is defined as follows.

Problem *(k competitive tracking with limited communications):* Under the Assumption, design a coordination control law for n redundant manipulators described by (6.5) to compete with each others such that the k redundant manipulators with their end-effectors being nearest from the moving target stay active for tracking while the others are deactivated and keep unmoved.

6.3 Dynamic Task Allocation with Limited Communications

In this section, we present a distributed competition control law for dynamic task allocation in the coordination of multiple redundant manipulators for path tracking with limited communications.

A k-WTA neural network model is presented in [15], which is described by

$$\frac{dz}{dt} = -\lambda \left(\sum_{i=1}^{n} w_i - k \right), \tag{6.6}$$

$$w_i = g_{\Omega i} \left(z + \frac{v_i}{2a} \right). \tag{6.7}$$

where $z \in \mathbb{R}$ is a auxiliary variable, $\lambda > 0$ is used to scale the convergence rate, v_i denotes the ith element of input vector v, w_i stands for the ith element of output vector $w \in \{0, 1\}^n$, a is a constant being enough small, $g_{\Omega i}(\cdot)$, as the ith element of projection function $g_{\Omega}(\cdot)$, is defined by

$$g_{\Omega i}(z + \frac{v_i}{2a}) = \begin{cases} 1, & \text{if } z + \frac{v_i}{2a} > 1 \\ z + \frac{v_i}{2a}, & \text{if } 0 \le z + \frac{v_i}{2a} \le 1 \\ 0, & \text{if } z + \frac{v_i}{2a} < 0. \end{cases} \tag{6.8}$$

Under the conditions that the kth largest element in v denoted by $\bar{v}_k$ is strictly larger than the $k + 1$th largest one denoted by $\bar{v}_{k+1}$, and that the constant parameter a satisfies $a \le 0.5(\bar{v}_k - \bar{v}_{k+1})$, it has been proven in [15] that the above k-WTA neural network model can be used for solving the following k-WTA problem:

$$w_i = \phi(v_i) = \begin{cases} 1, & \text{if } v_i \in \{k\text{largest elements of } v\} \\ 0, & \text{otherwise.} \end{cases} \tag{6.9}$$

In addition, we define the WTA index as

$$v_i = -\varepsilon_i = -\|r_d - f_i(\theta_i)\|_2^2/2. \tag{6.10}$$

The movement control for the ith redundant manipulator can be formulated as

$$\dot{\theta}_i = w_i c_0 \tau_i = w_i c_0 J_i^{\mathrm{T}}(r_{\mathrm{d}} - f_i(\theta_i)), \tag{6.11}$$

where τ_i is defined as $\partial v_i / \partial \theta_i = J_i^{\mathrm{T}}(r_{\mathrm{d}} - f_i(\theta_i))$. It can be readily concluded from (6.11) that, for the situation of $w_i = 0$, the ith manipulator is deactivated and unmoved, and that, for the situation of $w_i = 1$, the end-effector of ith manipulator executes the path-tracking task.

Substituting (6.7) into (6.6) and (6.11), we obtain

$$
\begin{cases}
\dot{\theta}_i = g_{\Omega i}\left(z + \frac{v_i}{2a}\right) c_0 \tau_i, \\
\dot{z} = -\lambda \left(\sum_{i=1}^{n} g_{\Omega i}(z + \frac{v_i}{2a}) - k\right).
\end{cases}
\tag{6.12}
$$

Combining the control input of all manipulators in the group, the coordination control law for executing the path-tracking task can be written into a compact form:

$$
\begin{cases}
\dot{\theta} = g_{\Omega}\left(z I_{2n} + \frac{v}{2a} \otimes I_2\right) c_0 \Phi, \\
\dot{z} = -\lambda \left(I_n^{\mathrm{T}} g_{\Omega}\left(z I_n + \frac{v}{2a}\right) - k\right),
\end{cases}
\tag{6.13}
$$

where $\theta = [\theta_1, \ldots, \theta_n]^{\mathrm{T}}$; $\Phi = [\tau_1, \ldots, \tau_n]^{\mathrm{T}}$; $\otimes$ is the Kronecker product; I_{2n}, I_n and I_2 denote vectors composed of $2n$, n and 2 elements with each element being 1, respectively.

It can be observed from (6.12) that the summation term $\sum_{i=1}^{n} g_{\Omega i}(z + v_i/(2a))$ requires information from every manipulator in the group and thus, this term is the obstacle to distributing the centralized coordination model. A consensus estimator is presented in [24] to decentralize an averaging operation, which allows n agents (i.e., manipulators in this chapter) with each of them measuring the dynamic term $g_{\Omega i}(z + \frac{v_i}{2a})$ and computes an approximation of $\bar{g}_{\Omega i}(z + \frac{v_i}{2a}) = \frac{1}{n} \sum_{i=1}^{n} g_{\Omega i}(z + \frac{v_i}{2a})$ using only limited communication. With the aid of the consensus estimator presented in [24], a manipulator is able to estimate the average of filter inputs by running the following protocol:

$$
\begin{cases}
\dot{\rho}_i = -\gamma \sum_{j \in \mathrm{N}(i)} A_{ij}(\rho_i - \rho_j) - \gamma(\rho_i - g_{\Omega i}(z + \frac{v_i}{2a})) \\
\quad -\gamma \sum_{j \in \mathrm{N}(i)} A_{ij}(\rho_i - \rho_j), \\
\dot{\rho}_i = \sum_{j \in \mathrm{N}(i)} A_{ij}(\rho_i - \rho_j),
\end{cases}
\tag{6.14}
$$

where ρ_i is the estimate of $\frac{1}{n} \sum_{i=1}^{n} g_{\Omega i}(z + \frac{v_i}{2a})$; $\mathrm{N}(i)$ denotes the neighbor set of the ith manipulator on the communication graph defined in the Assumption; ρ_i is scalar state maintained by the ith manipulator; A_{ij} is a positive constant for $j \in \mathrm{N}(i)$ and satisfies $A_{ij} = A_{ji}$; γ is a positive constant used to scale the convergence rate. Besides, it is worth mentioning that, for $j \notin \mathrm{N}(i)$, $A_{ij} = A_{ji} = 0$. By running (6.14) on every manipulator, ρ_i is able to track the average of inputs, i.e., $\sum_{i=1}^{n} w_i/n$ or $\sum_{i=1}^{n} g_{\Omega i}(z + v_i/(2a))/n$.

Thus, the term $\sum_{i=1}^{n} g_{\Omega i}(z+v_i/(2a))$ in (6.12) can be replaced with the distributed filter (6.14). Then, we have

$$
\begin{cases}
\dot{\rho}_i = -\gamma \sum_{j \in N(i)} A_{ij}(\rho_i - \rho_j) - \gamma(\rho_i - g_{\Omega i}(z + \frac{v_i}{2a})) \\
\quad\quad -\gamma \sum_{j \in N(i)} A_{ij}(\rho_i - \rho_j), \\
\dot{\rho}_i = \sum_{j \in N(i)} A_{ij}(\rho_i - \rho_j) \\
\dot{\theta}_i = g_{\Omega i}(z + \frac{v_i}{2a})c_0\tau_i, \\
\dot{z} = -\lambda(n\rho_i - k).
\end{cases}
\tag{6.15}
$$

Combining the control inputs of all redundant manipulators in the group, the coordination control law for executing the path-tracking task in the situation of limited communications can be written into a compact form:

$$
\begin{cases}
\dot{\rho} = -\gamma L\rho - \gamma(\rho - w) - \gamma L \int_{t_0}^{t} L\rho \mathrm{d}t, \\
\dot{\theta} = g_{\Omega}(zI_{2n} + \frac{v}{2a} \otimes I_2)c_0\Phi, \\
\dot{z} = -\lambda(I_n^T\rho - k),
\end{cases}
\tag{6.16}
$$

where t_0 denotes the initial time instant; Laplacian matrix $L = \mathrm{diag}(AI_n) - A$ with $\mathrm{diag}(AI_n)$ being the diagonal matrix whose n diagonal entries are the n elements of the vector AI_n with the ijth element of matrix A being A_{ij}.

Remark 6.1 A stable distributed coordination control law (6.16) for the dynamic task allocation in multiple manipulators coordination for executing the path-tracking task can be expected if the consensus filter (6.14) runs fast enough relative to the centralized coordination control law (6.13). That is to say, if the parameter γ is large enough relative to λ, then we expect the resulting dynamics to converge semiglobally.

It is difficult to conduct rigorous analysis on the convergence and stability of coordination control law (6.16). Therefore, for simplicity, we have the following theorem to analyze the coordination control law (6.16) based on its original model (6.13) and Remark 6.1.

Theorem 1 *For a group of n redundant manipulators described by (6.1) and the coordination control law (6.13), exactly k redundant manipulators with the minimum distance are activated and their end-effectors move towards to the moving target with time.*

Proof Define

$$
V_0 = \lambda \left[\sum_{i=1}^{n} h\left(z + \frac{v_i}{2a}\right) - kz) \right],
$$

with

$$h(x) = \begin{cases} 0, & \text{if } x < 0 \\ \frac{x^2}{2}, & \text{if } 0 \le x \le 1 \\ -\frac{1}{2} + x, & \text{if } x > 1. \end{cases}$$

For the properties of $h(x)$, we have the following results.

- $\sum_{i=1}^{n} \left[h(z + \frac{v_i}{2a}) - \frac{k(z + v_i/2a)}{n} \right]$ is lower bounded. Note that

$$h(x) - \frac{kx}{n} = \begin{cases} -\frac{kx}{n}, & \text{if } x < 0 \\ \frac{x^2}{2} - \frac{kx}{n}, & \text{if } 0 \le x \le 1 \\ -\frac{1}{2} + \frac{n-k}{n}x, & \text{if } x > 1. \end{cases}$$

It can be concluded that

$$h(x) - \frac{kx}{n} = \begin{cases} -\frac{kx}{n} \ge 0, & \text{if } x < 0 \\ \frac{x^2}{2} - \frac{kx}{n} \ge -\frac{k^2}{2n^2}, & \text{if } 0 \le x \le 1 \\ -\frac{1}{2} + \frac{n-k}{n}x \ge \frac{n-k}{n} - \frac{1}{2}, & \text{if } x > 1. \end{cases}$$

Therefore, $\sum_{i=1}^{n} \left[h(z + \frac{v_i}{2a}) - \frac{k(z + v_i/2a)}{n} \right]$ is lower bounded.
- It can be obtained that $\partial h(x)/\partial x = g_\Omega(x)$.

In addition, let

$$\Psi = \frac{2a}{c_0 \lambda} V_0 + \frac{1}{c_0} \sum_{i=1}^{n} \varepsilon_i.$$

We have the following results on the properties of Ψ,

- Ψ is lower bounded. We have

$$\Psi = \frac{2a}{c_0} \sum_{i=1}^{n} \left[h(z + \frac{v_i}{2a}) - \frac{k}{n}(z + \frac{v_i}{2a}) \right] + \frac{n-k}{c_0 n} \sum_{i=1}^{n} (-v_i).$$

As proven above, $\sum_{i=1}^{n} \left[h(z + \frac{v_i}{2a}) - \frac{k(z+v_i/2a)}{n} \right]$ is lower bounded. Besides, ε_i is also lower bounded. Therefore, Ψ is lower bounded.
- For $\dot{\Psi}$, we have the following results.

$$\dot{\Psi} = \left(\frac{\partial \Psi}{\partial z} \right)^{\mathrm{T}} \dot{z} + \sum_{i=1}^{n} \left(\frac{\partial \Psi}{\partial v_i} \right)^{\mathrm{T}} \dot{v}_i,$$

in which, we have

$$\frac{\partial \Psi}{\partial z} = \frac{2a}{c_0} \sum_{i=1}^{n} \left(g_{\Omega i} \left(z + \frac{v_i}{2a} \right) - \frac{k}{n} \right) = \frac{2a}{c_0} \sum_{i=1}^{n} \left(w_i - \frac{k}{n} \right) = \frac{2a}{c_0} \sum_{i=1}^{n} w_i - k,$$

$$\frac{\partial \Psi}{\partial v_i} = \frac{2a}{c_0} \left(g_{\Omega i} \left(z + \frac{v_i}{2a} \right) \frac{1}{2a} - \frac{k}{n} \frac{1}{2a} \right) + \frac{k}{c_0 n} - \frac{1}{c_0} = \frac{1}{c_0} \left(g_{\Omega i} \left(z + \frac{v_i}{2a} \right) - \frac{1}{c_0} \right).$$

and

$$\dot{v}_i = c_0 g_{\Omega i} (z + \frac{v_i}{2a}) \| J_i^{\mathrm{T}} (r_{\mathrm{d}} - f_i(\theta_i)) \|_2^2,$$

In addition, we further have

$$\left(\frac{\partial \Psi}{\partial z} \right)^{\mathrm{T}} \dot{z} = -\lambda \frac{2a}{c_0} \left(\sum_{i=1}^{n} w_i - k \right)^{\mathrm{T}} \left(\sum_{i=1}^{n} w_i - k \right) = -\lambda \frac{2a}{c_0} \left(\sum_{i=1}^{n} w_i - k \right)^2 \le 0,$$

and

$$\left(\frac{\partial \Psi}{\partial v_i} \right)^{\mathrm{T}} \dot{v}_i = \left(g_{\Omega i} \left(z + \frac{v_i}{2a} \right) - 1 \right) g_{\Omega i} \left(z + \frac{v_i}{2a} \right) \| J_i^{\mathrm{T}} (r_{\mathrm{d}} - f_i(\theta_i)) \|_2^2.$$

It can be concluded from (6.8) that

$$\left(g_{\Omega i} \left(z + \frac{v_i}{2a} \right) - 1 \right) \le 0$$

$$g_{\Omega i} \left(z + \frac{v_i}{2a} \right) \ge 0.$$

Then, we have

$$\left(\frac{\partial \Psi}{\partial v_i} \right)^{\mathrm{T}} \dot{v}_i \le 0,$$

with $=$ holding for $g_{\Omega i} \left(z + \frac{v_i}{2a} \right) = 1$ or $g_{\Omega i} \left(z + \frac{v_i}{2a} \right) = 0$. In addition, it can be obtained readily that $\dot{\Psi} \le 0$.

Using LaSalle's principle [25] (also known as the LaSalle's invariance principle, a criterion for the asymptotic stability of an autonomous dynamical system), let $\dot{\Psi} = 0$, and we have, $\forall i$,

$$\left(g_{\Omega i} \left(z + \frac{v_i}{2a} \right) - 1 \right) g_{\Omega i} \left(z + \frac{v_i}{2a} \right) \| r_{\mathrm{d}} - f_i(\theta_i) \|_2^2 = 0, \qquad (6.17)$$

and

$$\sum_{i=1}^{n} w_i = k. \tag{6.18}$$

For the above two cases, we have the following analysis results.

- As to the case of (6.17), we have the following three subcases.

  - Subcase 1. $g_{\Omega i}(z + \frac{v_i}{2a}) = 1 \Rightarrow w_i = 1 \Rightarrow \dot{\theta}_i = c_0 \tau_i$, and finally, we can have $f_i(\theta_i) \to r_d$ as $t \to \infty$.
  - Subcase 2. $g_{\Omega i}(z + \frac{v_i}{2a}) = 0 \Rightarrow w_i = 0 \Rightarrow \dot{\theta}_i = 0$, and finally, we can have that the ith redundant manipulator is deactivated and $f_i(\theta_i)$ is unmoved.
  - Subcase 3. $r_d - f_i(\theta_i) = 0 \Rightarrow f_i(\theta_i) = r_d$.

- As to the case of (6.18), we have

$$\sum_{i=1}^{n} w_i = k = \sum_{i=1}^{n} g_{\Omega i}(z + \frac{v_i}{2a}).$$

Reordering v_i for $i = 1, 2, \ldots, n$ as $v_1^* \geq v_2^* \geq \cdots v_n^*$. Then, we have $g_{\Omega i}(z + \frac{v_1^*}{2a}) \geq g_{\Omega i}(z + \frac{v_2^*}{2a}) \geq \cdots \geq g_{\Omega i}(z + \frac{v_n^*}{2a})$.

With $n1 + n2 + n3 = n$, assume that $g_{\Omega i}(z + \frac{v_1^*}{2a}) = g_{\Omega i}(z + \frac{v_2^*}{2a}) = \cdots = g_{\Omega i}(z + \frac{v_{n1}^*}{2a}) = 1$, that the values of $g_{\Omega i}(z + \frac{v_{n1+1}^*}{2a})$, $g_{\Omega i}(z + \frac{v_{n1+2}^*}{2a}), \ldots, g_{\Omega i}(z + \frac{v_{n1+n2}^*}{2a}) \in (0, 1)$, and that $g_{\Omega i}(z + \frac{v_{n1+n2+1}^*}{2a}) = g_{\Omega i}(z + \frac{v_{n1+n2+2}^*}{2a}) = \cdots = g_{\Omega i}(z + \frac{v_{n1+n2+n3}^*}{2a}) = 0$. We further have the following three subcases.

- Subcase 1. For v_i^* with $i \in \{1, \ldots, n1\}$, we have $w_i = 1 \Rightarrow \dot{\theta}_i = c_0 \tau_i$, and finally, we can have $f_i(\theta_i) \to r_d$ as $t \to \infty$.
- Subcase 2. For v_i^* with $i \in \{n1 + 1, \ldots, n1 + n2\}$, we have $w_i > 0$, and finally, we can have $f_i(\theta_i) \to r_d$ as $t \to \infty$.
- Subcase 3. For v_i^* with $i \in \{n1 + n2 + 1, \ldots, n1 + n2 + n3\}$, we have $w_i = 0 \Rightarrow \dot{\theta}_i = 0$, and finally, we can have that the ith redundant manipulator is deactivated and $f_i(\theta_i)$ is unmoved.

For the subcases 1 and 2, v_i approaches maximum for both cases and thus $g_{\Omega i}(z + \frac{v_{n1+1}^*}{2a})$ reaches the same value for them. In addition, we have

$$k = \sum_{i=1}^{n} g_{\Omega i}(z + \frac{v_i^*}{2a}) = n1 + \sum_{i=n1+1}^{n1+n2} g_{\Omega i}(z + \frac{v_i^*}{2a}).$$

As stated above, for $i \in \{n1 + 1, \ldots, n1 + n2\}$,

$$g_{\Omega i}(z + \frac{v_i^*}{2a}) = 1.$$

Thus, $k = n1 + n2$ and all of them are winners in the sense that $g_{\Omega i}(z + \frac{v_i}{2a}) = 1$.

Therefore, for a group of n redundant manipulators described by (6.1) and the coordination control law (6.13), exactly k redundant manipulators with the minimum distance are activated and their end-effectors move towards to the moving target with time. The proof is thus completed.   □

We have the following five remarks on the proposed distributed coordination control law (6.16) applied to the dynamic task allocation in distributed coordination of multiple redundant robot manipulators for path-tracking with limited communications.

Remark 6.2 To incorporate the constraint of joint-velocity limits into the proposed distributed coordination control law (6.16), a simple and direct method is to employ the following saturation function with $\dot{\theta}_i^+$ and $\dot{\theta}_i^-$ denoting the upper and lower limits of the joint-velocity of the ith redundant manipulator, respectively:

$$\mathbb{S}(\dot{\theta}_i) = \begin{cases} \dot{\theta}_i^+, & \text{if } \dot{\theta}_i > \dot{\theta}_i^+ \\ \dot{\theta}_i, & \text{if } \dot{\theta}_i^- \leq \dot{\theta}_i \leq \dot{\theta}_i^+ \\ \dot{\theta}_i^-, & \text{if } \dot{\theta}_i < \dot{\theta}_i^-. \end{cases}$$

Remark 6.3 To handle the range limits of joint-angles and joint-velocities simultaneously, a conversion technique based on techniques investigated in [9] is provided in this remark. Since the proposed distributed coordination control law (6.16) is solved at the joint-velocity level, the joint physical limits can be converted into a bound constraint in terms of joint velocity $\dot{\theta}_i$. The new bounds can, thus, be written as $\xi_i^- \leq \dot{\theta}_i \leq \xi_i^+$, with the jth elements of ξ_i^- and ξ_i^+ being defined, respectively, as

$$\xi_{ij}^- = \max\{\kappa_i(\theta_{ij}^- - \theta_{ij}), \dot{\theta}_{ij}^-\},$$
$$\xi_{ij}^+ = \min\{\kappa_i(\theta_{ij}^+ - \theta_{ij}), \dot{\theta}_{ij}^+\},$$

where κ_i is the scaling factor for the ith redundant manipulator used to determine the deceleration magnitude when a joint approaches its limits; $\theta_{ij}, \theta_{ij}^-, \theta_{ij}^+, \dot{\theta}_{ij}^-, \dot{\theta}_{ij}^+$ denote the jth element of $\theta_i, \theta_i^-, \theta_i^+, \dot{\theta}_i^-, \dot{\theta}_i^+$, respectively. In mathematics, κ_i should be greater than or equal to $2\max_{1\leq j\leq p}\{(\dot{\theta}_{ij}^+/(\theta_{ij}^+ - \theta_{ij}), -\dot{\theta}_{ij}^-/(\theta_{ij}^+ - \theta_{ij}))$ with p denoting the number of joints for the ith manipulator. Therefore, a new saturation function can be written as

$$\tilde{\mathbb{S}}(\dot{\theta}_i) = \begin{cases} \xi_i^+, & \text{if } \dot{\theta}_i > \xi_i^+ \\ \dot{\theta}_i, & \text{if } \xi_i^- \leq \dot{\theta}_i \leq \xi_i^+ \\ \xi_i^-, & \text{if } \dot{\theta}_i < \xi_i^-. \end{cases}$$

Remark 6.4 In view of the facts that the end-effector of a redundant manipulator has limited operation plane and that the manipulator with the nearest distance from its end-effector to the desired target may fall into the singularity situation, we present a criterion to avoid such a situation. That is, if the sum of the absolute values of the second joint angle to the pth joint angle is less than 0.05, then the corresponding v_i is adjusted to $v_i - \varsigma$, where ς denotes a positive parameter with large enough value. It can be readily deduced that the $k + 1$th redundant manipulator will win the competition and be activated to execute the path-tracking task after such an adjustment.

Remark 6.5 For the situation of $v_i = v_{i+1}$, it is an equilibrium point of the proposed distributed coordination control law (6.16). However, it is not a stable equilibrium point of the proposed control law. Note that, there are inevitable perturbations, in the form of noise from sensors or disturbance from actuators, under real-world conditions. Any small perturbations can trigger a non-reversible process to distinguish v_i from v_{i+1}.

6.4 Illustrative Example

In this section, computer simulations are conducted based on 12 redundant manipulators executing a path-tracking task in a competition manner to illustrate the effectiveness of the distributed dynamic coordination control law (6.16) with limited communications, where the desired path is the trajectory of a moving target. In the examples, we choose $a = 0.1$, $c_0 = 30$, $\lambda = 10$, $\gamma = 10^5$, $\varsigma = 10000$, $p = 4$ with each link length in each manipulator is 1 m and the task duration is 20 s. In addition, A_{ij} is set as

$$A_{ij} = \begin{cases} 1, \text{ if } |i - j| \leq 1 \\ 0, \text{ otherwise.} \end{cases}$$

Besides, the initial values of the rest parameters are set as 0. The distributed coordination control law (6.16) with joint-velocity limitation is investigated with $k = 1$. The corresponding simulation results are illustrated in Fig. 6.2 through Fig. 6.5.

Specifically, Fig. 6.2 shows the entire process of the path-tracking task handled by distributed coordination control law (6.16), where the initial position of the base of each redundant manipulator is randomly placed with the initial joint angle of each manipulator being $[\pi/2, \pi/4, 0, \pi/6]^T$. In addition, the initial values of distributed coordination control law (6.16) are randomly generated. As shown in Fig. 6.2a, at the initial time, the moving target is located at around $(0, 4)$ and the redundant manipulator with its end-effector nearest to the moving target (in terms of the shortest distance to the target) is activated to execute the path-tracking task. As the moving target approaches to the end-effector of one of the losers, the winner at the initial time fails in the competition and becomes a loser afterward. As a continuator, the

Fig. 6.2 With $k = 1$ and $n = 12$, path-tracking performance synthesized by distributed coordination control law (6.16), where each *blue circle* in subfigure (**a**) denotes the initial location of the end-effector of a redundant manipulator. **a** Desired path generated by the moving target and tracking trajectories of different redundant manipulators. **b** Profiles of joint-velocity of all the redundant manipulator. **c** The output of k-WTA network

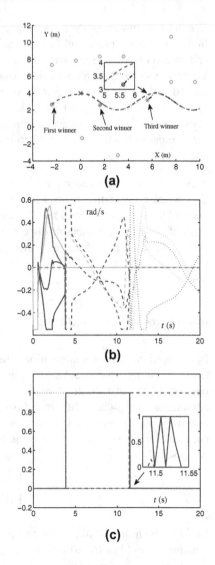

tracking task is allocated dynamically to the new winner labeled as the second winner and the latter's end-effector begins to track the desired path generated by the moving target. Then, as illustrated in the figure, the end-effector of the third winner redundant manipulator relays to execute the task. The corresponding profiles of joint-velocity of all redundant manipulators are visualized in Fig. 6.2b. It can be readily observed from the figure that, corresponding to the coordination tracing performance shown in Fig. 6.2a, at each time instant, only four joints move while the value of the rest ones remain zero. It is worth pointing out that, as shown in Fig. 6.2b, several joint velocities reach lower limit $\dot{\theta}^-$ and upper limit $\dot{\theta}^+$, but never violate them, which demonstrates that the constraint of joint-velocity limits presented in Remark 6.2 works effectively.

Fig. 6.3 Profiles of joint-angle and joint-velocity of the first winner redundant manipulator as well as the corresponding motion trajectories synthesized by distributed coordination control law (6.16) with limited communications. **a** Profiles of joint-angle. **b** Profiles of joint-velocity. **c** Motion trajectories

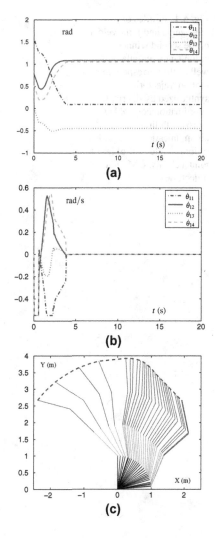

It can be found in Fig. 6.2c that, starting with randomly generated initial state, the outputs of the k-WTA network rapidly converge to the correct results, and the outputs change rapidly when the target approaches to the new winner. In addition, as shown in the square-marked area, oscillation appears during the change of output, which is mainly because of the estimation on $\sum_{i=1}^{n} g_{\Omega i}(z + v_i/(2a))$ via consensus filter (6.14).

To observe the dynamic task allocation in multiple redundant manipulator coordination for path-tracking in detail, simulation results on different phases of the path-tracking task are illustrated in Fig. 6.3 through Fig. 6.5. It can be seen from Fig. 6.3a that, starting from the given initial state, the four joints of the first winner redundant manipulator work well to track the desired path generated by the moving

Fig. 6.4 Profiles of
joint-angle and joint-velocity
of the second winner
redundant manipulator as
well as the corresponding
motion trajectories
synthesized by distributed
coordination control law
(6.16) with limited
communications. **a** Profiles
of joint-angle. **b** Profiles of
joint-velocity. **c** Motion
trajectories

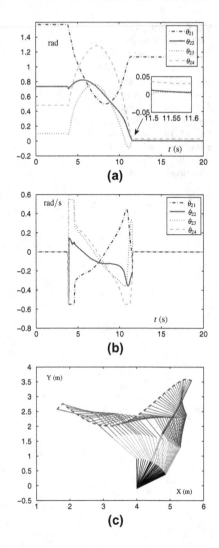

(a)

(b)

(c)

target. In addition, these joints stay unmoved after 4 s, which means that the path-tracking task is being delivered to the second winner redundant manipulator. The corresponding joint velocities of the first winner redundant manipulator are shown in Fig. 6.3b, which are kept in the joint-velocity bound and remain 0 after 4 s. The motion trajectories of the first winner redundant manipulator are shown in Fig. 6.3c, from which we can see how the manipulator tracks the desired path with four joints. It is worth pointing out here that, as shown in the square-marked area in Fig. 6.4a, the criterion presented in Remark 6.4 is satisfied and thus the corresponding adjustment is activated to avoid the possible singularity. Therefore, the path-tracking task is allocated to the third winner manipulator. The simulation results on the second

Fig. 6.5 Profiles of
joint-angle and joint-velocity
of the third winner redundant
manipulator as well as the
corresponding motion
trajectories synthesized by
distributed coordination
control law (6.16) with
limited communications.
a Profiles of joint-angle.
b Profiles of joint-velocity.
c Motion trajectories

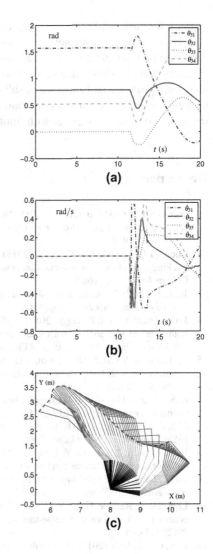

and the third phases are similar to that of the first phase, and thus the corresponding descriptions are omitted. These simulation results verify the effectiveness of the proposed distributed coordination control law (6.16) with limited communications.

6.5 Summary

In this chapter, a new coordination control law of multiple redundant manipulators has been defined for task allocation in executing a path-tracking task, in which only the winners allocated with the task are activated with their end-effectors commended to track the task while the rest ones keep unmoved. We have proposed a distributed coordination control law with limited communications and with the aid of a distributed consensus filter. The stability of the distributed control has been proved in theory. Finally, illustrative simulation examples based on redundant manipulators have been provided and analyzed to substantiate the efficacy of the proposed distributed coordination control law for dynamic task allocation in executing path-tracking tasks in a competition manner with limited communications.

References

1. Wen G, Chen CLP, Liu Y, Liu Z (2015) Neural-network-based adaptive leader-following consensus control for second-order non-linear multi-agent systems. IET Control Theory Appl 9(13):1927–1934
2. Chen CLP, Wen G, Liu Y, Liu Z (2015) Observer-based adaptive backstepping consensus tracking control for high-order nonlinear semi-strict-feedback multiagent systems. IEEE Trans Cybern 46(7):1591–1601
3. Jin L, Li S (2017) Distributed task allocation of multiple robots: A control perspective. IEEE Trans Syst Man Cybern Syst pp(99):1–9
4. Jin L, Zhang Y, Li S, Zhang Y (2016) Modified ZNN for time-varying quadratic programming with inherent tolerance to noises and its application to kinematic redundancy resolution of robot manipulators. IEEE Trans Ind Electron 63(11):6978–6988
5. Jin L, Zhang Y (2015) Discrete-time Zhang neural network for online time-varying nonlinear optimization with application to manipulator motion generation. IEEE Trans Neural Netw Learn Syst 27(6):1525–1531
6. Li S, Kong R, Guo Y (2014) Cooperative distributed source seeking by multiple robots: Algorithms and experiments. IEEE/ASME Trans Mech 19(6):1810–1820
7. Li S, Zhang Y, Jin L (2016) Kinematic control of redundant manipulators using neural networks. IEEE Trans Neural Netw Learn Syst. doi:10.1109/TNNLS.2016.2574363 (In Press)
8. Li S, Chen S, Liu B, Li Y, Liang Y (2013) Decentralized kinematic control of a class of collaborative redundant manipulators via recurrent neural networks. Neurocomputing 91:1–10
9. Jin L, Zhang Y (2015) G2-type SRMPC scheme for synchronous manipulation of two redundant robot arms. IEEE Trans Cybern 45(2):153–164
10. Li S, He J, Rafique U, Li Y (2017) Distributed recurrent neural networks for cooperative control of manipulators: A game-theoretic perspective. IEEE Trans Neural Netw Learn Syst 28(2):415–426
11. Li S, Cui H, Li Y (2013) Decentralized control of collaborative redundant manipulators with partial command coverage via locally connected recurrent neural networks. Neural Comput Appl 23(1):1051–1060
12. Li S, Zhou M, Luo X, You Z (2017) Distributed winner-take-all in dynamic networks. IEEE Trans Autom Control 62(2):577–589
13. Maass W (2000) On the computational power of winner-take-all. Neural Comput 12(11):2519–2535
14. Liu S, Wang J (2006) A simplified dual neural network for quadratic programming with its kwta application. IEEE Trans Neural Netw 17(6):1500–1510
15. Hu X, Wang J (2006) An improved dual neural network for solving a class of quadratic programming problems and its k-winners-take-all application. IEEE Trans Neural Netw 19(12):2022–2031
16. Li S, Liu B, Li Y (2013) Selective positive-negative feedback produces the winner-take-all competition in recurrent neural networks. IEEE Trans Neural Netw Learn Syst 24(2):301–309

17. Jin L, Li S, La H, Luo X (2017) Manipulability optimization of redundant manipulators using dynamic neural networks. IEEE Trans Ind Electron pp(99):1–10. doi:10.1109/TIE.2017.2674624 (In press)

18. Zhang Y, Li S (2017) Predictive suboptimal consensus of multiagent systems with nonlinear dynamics. IEEE Trans Syst Man Cybern Syst pp(99):1–11. doi:10.1109/TSMC.2017.2668440 (In press)

19. Jin L, Zhang Y, Qiu B (2016) Neural network-based discrete-time Z-type model of high accuracy in noisy environments for solving dynamic system of linear equations. Neural Comput Appl. doi:10.1007/s00521-016-2640-x (In press)

20. Li S, You Z, Guo H, Luo X, Zhao Z (2016) Inverse-free extreme learning machine with optimal information updating. IEEE Trans Cybern 46(5):1229–1241

21. Khan M, Li S, Wang Q, Shao Z (2016) CPS oriented control design for networked surveillance robots with multiple physical constraints. IEEE Trans Comput-Aided Des Integr Circuits Syst 35(5):778–791

22. Khan M, Li S, Wang Q, Shao Z (2016) Formation control and tracking for co-operative robots with non-holonomic constraints. J Intell Rob Syst 82(1):163–174

23. Jin L, Zhang Y (2016) Continuous and discrete Zhang dynamics for real-time varying nonlinear optimization. Numer Algorithm 73(1):115–140

24. Freeman R, Yang P, Lynch K (2006) Stability and convergence properties of dynamic average consensus estimators. In: Proceedings of IEEE CDC, pp 338–343

25. Hespanha J (2004) Uniform stability of switched linear systems: extensions of LaSalle's invariance principle. IEEE Trans Autom Control 49(4):470–482

Printed in the United States
By Bookmasters